Samuel Kneeland

**The Wonders of the Yosemite Valley**

And of California

Samuel Kneeland

**The Wonders of the Yosemite Valley**
*And of California*

ISBN/EAN: 9783744666725

Printed in Europe, USA, Canada, Australia, Japan

Cover: Foto ©berggeist007 / pixelio.de

More available books at **www.hansebooks.com**

# THE WONDERS OF THE YOSEMITE VALLEY,

### AND OF

## CALIFORNIA.

(v)

## Dedication.

---

(ix)

# PREFACE.

NORTH AMERICA is certainly a favored land in its magnificent scenery: in its White and Green Mountains, Adirondacks, Appalachians, Rocky Mountains and Sierra Nevada; in its great lakes; in its mighty rivers — the Mississippi, Missouri, Colorado, and their tributaries; in its cataracts — Niagara, Genesee, Trenton, Ithaca, Montmorenci, Minnehaha, and the grand cascades of the Yosemite Valley; in its boundless prairies, magnificent forests, and variety of the aspects of nature from the tropics to the arctic regions. If it be possible for grandeur of natural scenery alone to elevate the mind, the Americans should be a people of great ideas.

It is a fact of which comparatively few seem to be aware, that California, the land of gold, is also the land of wonders in scenery and in natural productions. To many of those who are cognizant of this fact, the distance from home, and the consequent fatigue and inconvenience of travel, appear as insurmountable obstacles. The first and the only difficulty in the journey to the Pacific is, to get started; that accomplished, with the comfortable cars, good food, easy bed, and other luxuries of the Pullman and Wagner palaces, the traveller of ordinary endurance and common-sense has only to take his ease and enjoy himself; if, to the above simple qualifications, he fortunately add a natural love of the picturesque, the grand, and the beautiful, I know of no journey on the face of the earth in which so much enjoyment can be crowded into a month's time.

In the lover of mountain scenery — even in one familiar with the Alps — the Rocky Mountains, and especially the Sierra Nevada, will excite a new and exquisite sensation. Such extent of grandeur is unparalleled in any mountains explored in civilized regions.

It does not require strong nerves, firm determination, nor great physical endurance, to make the trip to the Yosemite; and this magnificent scenery is easily within the reach of the invalid, male or female, who is not so hopelessly enfeebled as to forbid, under any circumstances, removal from home.

The beauties and wonders described in this book, however, are not presented for the benefit of the sick, but to the crowd of pleasure-seekers who make their annual visitations to Niagara, Newport, Saratoga, Cape May, and other centres of fashion, frivolity, foppery and folly. With half the expenditure of money and vital force thus thrown away, to the moral and physical deterioration of all

(xi)

concerned, the California trip, *via* the Pacific Railroad, may be thoroughly enjoyed. There is nothing in it to enfeeble, but everything to strengthen; the exhilarating mountain air, by day or by night, makes the lungs tingle with a sensation never experienced at the Eastern watering-places; the cool mountain-streams will prove a better tonic to the dyspeptic, than all the drugs he has swallowed. The brain of the student and the overworked merchant can here lie fallow amid scenes which, by their strange fascination, will drive from the memory all thought of books and ledgers; even the love of dress, and the pursuit of fashions, leave their votaries, as they take their seat in the saddle for the Valley or the Big Trees.

The absence of storms in the summer, the serenity of an unclouded sky, and a deliciously cool air, permit one to climb the mountains without the risk of getting wet, of being delayed by an avalanche, of falling into an ice-bound crevice, or of being enveloped in a thick mist, at a point noted for fine scenery, so provokingly common in Switzerland. Without danger, hardship, or even discomfort, and with a certainty of fine weather week after week, the California mountains invite you to their magnificent scenery.

Without any pretension to original discovery, or to the loftiness of style befitting so grand a subject, this volume is issued in the hope that the scenes recently visited by the writer may be more sought for by Eastern travellers; and that the order followed by him, and sketched imperfectly here, may serve in some measure as a useful guide to the grandeur of the Yosemite Valley, and to the other wonders of California.

S. K.

Boston, November, 1871.

THE

# WONDERS OF THE YOSEMITE VALLEY,

### AND OF

## CALIFORNIA.

### OMAHA TO SALT LAKE.

ON the east of the Rocky Mountains most of the great river sys-
tems descend very gradually, and pour their waters through the
Mississippi into the Gulf of Mexico, viz. : the Red, Arkansas, Rio
Grande, Platte, and Missouri ; while the Columbia and the Colorado
flow into the Pacific Ocean ; the former water lands of great luxuriance,
and thickly populated ; the latter flow through a sterile region,
hardly fit for the abode of man, yet with very grand scenery.

The profile of the Pacific Railroad, from Omaha to Sacramento,
1,775 miles, has four principal summits. 1. At Sherman, where the
Rocky Mountains (or Black Hills, so called) are crossed, 550 miles
from Omaha, 8,235 feet above the level of the sea, the highest point
in the world crossed by a railroad. 2. Aspen Summit, 385 miles from
Sherman, or 935 from Omaha, 7,463 feet high; also in the Rocky
Mountains, and the dividing ridge or continental rocky back-bone. 3.
In the Humboldt range, near Pequop, 310 miles from Aspen, or 1,245
from Omaha, 6,076 feet high. 4. In the Sierra Nevada, at Donner
Lake Pass, 425 miles from the Humboldt Summit, 1,670 from Omaha,
or 105 from Sacramento, 7,062 feet high; thence there is a descent
of 7,000 feet in 100 miles to Sacramento, very steep, and to the inex-
perienced traveller seemingly dangerous. The road from Cheyenne,
520 miles from Omaha, for 500 miles on a stretch, to the Wahsatch
Range in Utah, is more than 6,000 feet above the level of the sea;
from this to the Sierra crossing the average height is 5,000 feet, and
nowhere less than 4,000 ; whence it would be naturally supposed that
the road would be liable to become blocked by snow ; this, however,
is not the case, as the snow-sheds are a protection in the most exposed
regions of the Sierra Nevada.

The muddy Missouri River is crossed from Council Bluffs, Iowa, to
Omaha, Nebraska, and here the Union Pacific Railroad begins, 968
feet above the level of the sea, in the great valley drained by this
river and its tributaries. The ascent is so gentle that you do not perceive it, and yet when you have reached Cheyenne, you are 6,000

feet above the sea, ascending from 7 to 10 feet per mile. For 290 miles the road is along the main stream of the Platte river; along its banks are many fine farms and clumps of trees, and the sides of the track are variegated with beautiful flowers, among which are roses, larkspurs, and a fine white thistle. This was once a hunting-ground of the Indians for bison and antelope; the former is now rarely seen, but now and then an antelope will scamper away from the track, turning, when at a safe distance, to scrutinize the rushing train which disturbed him. This was also a portion of the road dangerous from Indians, as here they were accustomed to cross the plains, naturally hating the whites for expelling themselves and the game from their favorite haunts. Every station was once, of necessity, a fort; the frequent camps of mounted riflemen, and their presence as armed sentinels at the stations, showed that it was not yet considered safe to leave the road at the mercy of the hostile tribes.

The Platte River, though navigable, as the saying is, for nothing larger than a shingle, on account of its shallowness, sand-bars, and ever-shifting channel, drains an area of nearly 300,000 square miles; larger than all New England, New York, and Pennsylvania. It is, however, nature's highway for a railroad, and probably but for it, this Pacific Railroad might never have been built. The old emigrant road was along this river, and it can now be traced by the telegraph poles, skulls and bones of cattle, and now and then a grave, bearing testimony to the toil, privation, and death of the gold-seekers.

Columbus, 91 miles from Omaha, is, according to George Francis Train, the geographical centre of the United States, and, when he becomes President, will be a candidate for the government buildings. Grand Island, in Platte River, is about 80 miles long, and 4 wide; it is fertile, and well-wooded, and belongs to the United States. From 150 to 350 miles from Omaha you are within the range of the buffalo, but will probably see none, not even a track; this region is also infested by Indians, as shown by the fort-like and guarded stations; the cabins are low, covered with mud and turf, to render harmless the blazing arrows of the savages, and with loop-holes for defence. Here and there a sullen-looking fellow, indifferently armed, scowls at the passing or stopping train, but we saw no bands.

About 290 miles from Omaha you come to the north and south forks of the Platte River, and the railroad takes a westerly course between them. Soon Alkali is reached, in the alkali belt which extends for seventy or eighty miles westward; the soil and water are strongly impregnated with alkaline salts, the carbonates of the alkalies being so abundant that the earth may be used for raising bread. Here farms cease, and the country is of use only for grazing. Julesburg, 377 miles, was noted as a thieving, gambling place, as the terminus of the advancing road always was; shanties and tents were built in a night, and disappeared as if by magic, leaving nothing behind but a bad reputation, ruined chimneys, old boots, tin cans, and soiled cards. These harpy communities, when too bad, were occasionally exterminated by "Vigilance Committees." At Lodgepole, about 400 miles, the elevation is nearly 4,000 feet, and from this you perceive that you are ascending. About thirty-five miles beyond this is Prairie Dog City, so named because, for several hundred acres on both sides of the track, the earth is raised into little hillocks by these burrowing squirrel-like animals. Each occupant of a burrow sits erect on his hillock, scampering into his hole in the most ludicrous manner at the approach of danger; they are obliged to endure in their villages the presence of the burrowing owl, which lives in burrows deserted by, or forcibly taken from, the rodent by the lazy owl; they do not live together in the same hole, as far as I could observe or ascertain. This is to be the great pasture-land of the Continent, and was evidently once the bottom of a great lake or inland sea; the region extends for 700 miles north and south, on the east of the Rocky Mountains, and for 200 miles east and west, besides the innumerable valleys in the mountain ranges; there is an abundant supply of water in the valleys, and the nutritious grasses, nine to twelve inches high, are always green near the roots, however parched and cured at the top; cattle require no housing, and need only be prevented from straying; in winter the snow is so dry that it rolls off their backs, and does not chill them like our wet, clinging snows. Now that the railroad is here to bring the products to the Eastern markets, it is safe to say, that in a few years the untold wealth to be derived from raising cattle and sheep will bring to this region a large and vigorous population from the overcrowded Atlantic States.

At Cheyenne, Wyoming Territory, 516 miles, you are nearly 6,000 feet high; here the engines are doubled, and in thirty-three miles you ascend about 2,300 feet, or seventy feet in a mile. This place, where in 1867 there was only one house, has now several thousand inhabitants, and has the elements of a permanent increase, and will not fade away like most of the other railroad creations. It has its newspapers, schools, churches, manufactories, and extensive system of inland transportation, especially in connection with the rapidly-increasing mining interests of Colorado on the south. About fifteen miles from Cheyenne the grade becomes very steep, and you have fine views of the "Black Hills," the most eastern ranges of the Rocky Mountains. The scenery now becomes wild and rugged, and the masses of reddish felspathic rock are piled up in grand confusion. On arriving at the summit, at Sherman, named from the tallest general in our army, you are 8,235 feet above the sea, the highest point crossed by any railroad. The summit is bare, and the surrounding desolation grand and awful; the rocks and the road-bed are of a reddish color, which gives an unearthly aspect to the scenery. The air, after you get a few inspirations, is singularly exhilarating. This is 550 miles distant from Omaha, and affords a good view of Pike's and Long's Peaks, and other localities famous in the history of gold-seeking. The many cuts and snow fences show the physical and elemental difficulties which were encountered here.

Three miles from Sherman you come to Dale Creek, which is bridged by a framework structure 650 feet long, and 126 feet above the stream; the wooden trestles are laced strongly together, and present, at a distance, a very light and graceful structure. When you get upon it you shudder as you look down and see the stream a mere thread below, and feel the bridge quivering under the weight of the train to such a degree that water is thrown from barrels, placed there for putting out accidental fires; it is a relief to get upon *terra firma*, when every one draws a full breath, which is instinctively impossible during the transit. I fear that a terrible accident will some day occur here, as a fancied security from past immunity is apt to beget carelessness, and the bridge itself does not seem to me sufficiently strong for its peculiarly dangerous locality.

For more than twenty miles from Sherman the descent is so great that no steam is required, and the brakes are constantly applied; this distance brings us to Laramie Plain, the grade of which, however, is constantly changing. You pass numerous ridges of reddish sandstone, worn by the elements into the most fantastic shapes, as castles, forts, churches, chimneys, pyramids, etc., looking like a city changed to stone by the enchanter's wand; the general name of "buttes" is given to these, with a prefix according to the color or shape, as red, black, church buttes, etc.; some of these singular formations are 1,000 feet high, and in the distance are very interesting objects to the observant traveller.

The Laramie Plain has a fine grazing belt, sixty miles long by twenty wide, one of the finest stock-raising regions in the world, the alkaline quality of the soil and water making the growth of very nutritious grasses most luxuriant; this was once a grazing place for the buffalo, now rarely seen. When there is too much alkali, of course the soil is barren, and the water unfit for animals and man. This plain is 7,000 feet above the sea, and is much broken by the ranges of the Black Hills, which enclose, often, extensive and fine tablelands or "parks," sheltered from the wind, abundantly watered, with excellent timber and grass, and much mineral wealth, which will one day be a source of great prosperity. The distant peaks are here and there crested with snow, but you see no glaciers and eternal snows, as in the Alps, coming down into the valleys; at the base is generally nothing but a barren, treeless plain, plentifully stocked with the pale aromatic wild sage, and the home of the wild rabbit and antelope. It affords a good example of hundreds of miles of country which apparently can never be brought under cultivation, nor become fit for the residence of civilized man.

At Carbon, 656 miles, there is good supply of tertiary coal, the shaft being close to the track, the yield being 200 tons a day; the force which uplifted this table land broke up these coal-bearing strata, fortunately placing them so that they are easily workable, and exceedingly valuable where wood is so scarce.

At Creston, 740 miles, 7,000 feet high, is the dividing line of the continent, where streams flow easterly to the Gulf of Mexico, and westerly to the Pacific. Sage brush and alkali give the aspect of desolation to this central point of the grandest of our mountain ranges. Westward for thirty miles, the country is a barren alkaline desert, with a reddish tint, from salts of iron.

Green River Station, 846 miles, is so named from the river, which

flows into the Colorado; the water has a greenish hue, from the mi-
nute particles of the decomposed green slaty rocks which it washes;
it is a large, rapid stream, with good water, plentifully stocked with
trout. This region was evidently once the bed of a large lake, or
very wide river, and affords a great many moss agates. Here you
pass into Utah Territory.

Aspen, 940 miles, 7,463 feet high, the second highest point on the
Union Pacific Railroad, is so named from the tree of that name, which
grows on the sides of the mountains, spurs of the Uintah Range. It
will be noticed that there is an interval of about 100 miles between
the stations here mentioned, which will indicate to the reader what a
dreary and uninteresting region this is as a whole, with here and
there a place worthy of mention.

From Aspen the track descends through the cut made by the Weber
River through the Wahsatch Range, into Salt Lake Valley. At Wahsatch, 968 miles, after a good breakfast (and it may be here stated,
once for all, that the meals all along the route are excellent, at moderate price, and with plenty of time to eat), you plunge into the
famous Echo Cañon, flanked by the most magnificent scenery. Here
comes in a merry conductor, full of proverbs and wise sayings, ready
to do battle in words, (and for aught I know with fists,) for all sound
morality; he has a fair voice, and as he enters the car, preliminary to
taking the tickets, treats the passengers to a snatch of some song,
sacred or profane, which puts every body into good-humor, contrasting favorably with the boorishness so frequently met with in conduct-
ors who ride behind horse-flesh in our large cities. He invites you
to go to the rear or observation car, open above and on the sides,
affording an unobstructed view on all sides. The cars soon pass into
a tunnel, 770 feet long, approached by a long and rather shaky trestle-
work; here the jolly conductor (not a Mormon, as you at first suppose) cautions young people, and especially any who may be on their
bridal tours, to be sure that they select the right person before they
proceed to any little caresses suggested by the long, dark tunnel; according to his account, many ludicrous and provoking mistakes have
sometimes been revealed when the sudden darting of the train into
the daylight has shown the various attitudes of the passengers; from
failure to recognize the points of the compass in the light, moustaches
have been found under the wrong bonnets, and arms around the wrong
waists.

No words can describe the wild and grand scenery of the Echo
Cañon, at this pass narrowed to a mere chasm, between cliffs of reddish sandstone from 500 to 2,000 feet high, almost overhanging the
road, and carved by the elements into the most fantastic forms, whose
names and resemblances are pointed out by the communicative conductor. Excellent photographs for stereoscopic use have rendered
these scenes familiar to many, and, though giving but little idea of the
real grandeur, serve well to fix in the memory of those who have seen
them the momentary glimpses so rapidly taken from the rushing car.
The whistle of the locomotive starts a thousand echoes from the rocky
sides, chiefly on the right, the left sloping away to grassy meadows.

Here are seen the " Mormon Fortifications," 1,000 feet high, with the massive rocks still in place destined to have been rolled upon the United States troops sent in 1857 to attack this people; they were, however, never used. Echo Creek winds among the rocks, and is crossed thirty times in twenty-five miles. Occasionally is seen a small Mormon settlement, of long one-storied houses, surrounded by richly-cultivated fields; but the houses and fences are in bad repair, with slouchy, bearded men hanging about, and the women sad-eyed, homely, and poorly dressed —the tyranny of their creed impressing itself even on their external appearance.

Soon after leaving Echo City, you come to the "thousand mile tree," a vigorous evergreen, spared to mark the thousandth mile from Omaha — 2,650 miles from good old Boston. Then comes Weber Cañon, cut by the river of that name, more beautiful, if possible, than Echo Cañon, though only three miles long (Echo being eight); it is rendered more pleasing by the river which rushes by the side of the track, now a torrent, then a cascade, then a whirlpool, and then boiling rapids, according to the obstructions of its rocky bed and sides. In this, as in Echo Cañon, every second brings into view some new wonder or beauty. We can mention only two, both named from his Satanic Majesty, who seems to claim most that is sublime and awful, in the scenery west of the Rocky Mountains. The first is the " Devil's Slide," two vertical ridges of granite, on the left of the track, extending several hundred feet in height; the earth between the ridges, which are several yards apart, is covered with grass and flowers, rendering by contrast the gray rocky barriers very distinct. Passing this and Weber Station you come to the second, the " Devil's Gate," a narrow gorge through which the Weber River rushes, crossed by a bridge about fifty feet above the raging stream. You have no opportunity for fright or pleasure, as you are whirled along by the iron horse, which has no eye for scenery, and regards only time and space.

After passing through these fine cañons in the Wahsatch Range, you are in the Great Salt Lake Valley, though still, at Uintah Station, 4,550 feet above the sea. Eight miles more and you are in Ogden, the terminus of the Union Pacific, 1,032 miles from Omaha. This is a strictly Mormon town; the houses are widely scattered, but with fine gardens and orchards. Near the depot is the usual assortment of shanties, tents, and saloons. On the platform you will probably see Indians of the Shoshone tribe, in costumes partly civilized and partly savage; as a military hat with feather, pants, and coat, with dirty blanket, moccasins, and daubed with paint — with the unmistakable odor of the red man, indicating, to more senses than the eye, that frequent ablution is not one of his virtues.

## SALT LAKE AND THE CENTRAL PACIFIC RAILROAD.

AT Ogden the traveller takes the Utah Central Railroad, going south, and after a two hours' ride, of thirty-five miles, arrives in Salt Lake City, the temporal and spiritual head-quarters of President Brigham Young.

Surrounded as is Salt Lake Valley by lofty mountains, and cut off from civilization by a thousand miles of barren and almost impassable deserts, it is certainly a very remarkable instance of human industry, perseverance, and devotion to what they regarded as a divine precept, that the Mormons should have established such a prosperous community in this unpromising region. Salt Lake City was founded in 1847; it is situated in latitude 40 deg. 46 min. north, and longitude 112 deg. 6 min. west, at the base of the western slope of the Wahsatch Mountains, which you pass by the Echo and Weber Cañons.

The history of the rise and progress of this strange sect cannot be entered into here. Suffice it to say that it was organized in 1830 by Joseph Smith, in Ohio, under circumstances savoring strongly of delusion and fanaticism, if not of deception; it afterward removed to Jackson County, Missouri, and then to Nauvoo, Illinois, on the Mississippi. Persecuted for obvious reasons in 1844-45, the Mormons emigrated in 1846, under President Brigham Young, the successor of Joseph Smith, who, with his brother Hyrum, was murdered by a mob in 1844. Persecution followed them through Missouri and Iowa, and they reached Great Salt Lake, after much hardship, in the latter part of July, 1847, passing up the left bank of the Platte River, crossing at Fort Laramie, and over the mountains at the South Pass. In 1850 Utah was admitted into the Union as a Territory, though it applied for admission as a State under the name of "Deseret."

The city is four miles long and three miles wide, the streets at right angles to each other, 132 feet wide, with sidewalks of twenty feet. Each house is twenty feet from the line of the street, and is adorned usually by shrubbery and trees; water is brought from the mountains, and its fresh current runs freely through the gutters of the streets, with a sound and sight very refreshing on a hot day, as you walk along under the grateful shade, over the sidewalks. Most of the houses are of adobe, or sun-dried brick and wood, and a few of stone. The stores are well supplied with goods from the East, and with excellent articles of home manufacture, which the saints are, in a measure, forced to buy — the trade of the Gentiles being with each other and with strangers, and not much with the Mormons. The Mormon stores, generally co-operative, are known by the sign, "Holiness to the Lord." Church and State are closely united, the heads of the church being also the high civil officers. One-tenth of all a convert has, he pays, it is said, into the "Treasury of the Lord," and one-tenth of his yearly profits, and devotes one-tenth of his time for public works — resembling the system of tithing of the ancient Israelites. There is, besides, a tax on property for the revenue of the civil government. Outward prosperity, peace, and contentment, seem to reign ; poverty is unknown ; crime is rare, and severely punished, and the ordinary vices of our large cities are not seen, and most likely do not extensively exist — the one great evil, as we deem it, polygamy, swallows up all lesser vices by taking away one great incentive.

The Mormons regard their prosperity as a sign of the favor of heaven ; but outsiders more truly ascribe it to their industry, discipline, and concentration of energies on one purpose. Whatever may be thought of their religious views and consequent practices, they are undoubtedly sincere. The President is a man of remarkably clear mind and sound sense, and with great executive ability, equal to his responsible position ; sincere and active in everything which he considers good for the moral, intellectual, and material elevation of his people, whose confidence he fully enjoys. He is of commanding appearance, affable to strangers, and impresses you with the idea of strength, firmness, and resolution, which indeed are required to keep this anomalous community from falling to pieces by the slow but continual sapping of its foundation-tenets by the encroachments of Eastern principles.

The "spiritual-wife" system, which now seems tottering to its fall, was not an original tenet of the Mormon creed, forming no part of the teachings of its founders ; and probably would long since have met the fate deserved by such an abomination, had it not been in great measure kept out of public sight by the remoteness and isolation of this people. Even now, when public indignation is aroused for its extinction, the problem is a difficult one to solve in a way which shall punish or restrain the guilty ones in high places, without causing unmerited suffering to the deluded wives and innocent children.

I have before me the "Third Annual Catalogue" of the "University of Deseret," in Salt Lake City, for the years 1870–71. It contains the names of 580 pupils : 286 males, and 294 females, with those of 13 instructors. The courses of instruction in the classics, in the sciences, and in the normal studies, will compare favorably with those of our Eastern colleges, and seem admirably adapted to prepare the way for a better state of things, evidently now approaching rapidly, and to develop the great natural resources of this country. With a fertile soil, healthy climate, and inexhaustible mineral wealth, this land of beauty and grandeur must soon be the pasture and the mine, as it is the highway of the nation. Time only can solve the questions of statesmanship, civil polity, religion, and morality, presented by this singular community, whose centre is at Salt Lake City. When the iron will which rules this people ceases to exert its influence, the Mormon system will doubtless crumble away before the advancing tide of Eastern civilization, now so rapidly surrounding and permeating it by means of the Pacific Railroad ; yet, whether its life be long or short, this sect has made a pathway and a stopping-place for the westward march of the nation, and thus, involuntarily, have greatly advanced the progress of humanity. The city is beautifully situated, and, as seen from the surrounding hills, its so-called "Valley of the Jordan" is a perfect garden in the wilderness. With and without irrigation the crops are fine, and the fruit is excellent; the grasshoppers are a great plague, and sometimes so utterly destroy a growing crop as to require planting even a third time. Camp Douglass overlooks the city, and, in case of need, could soon shell out an enemy. The valley was evidently once the bottom of an inland sea, as proved by the terraces, which can be traced for miles along the sides of the mountains, indicating former levels of the water; it contains over 1,100 square miles, with much fine grazing, as well as cultivated, land. Mormon industry has shown that reclaimed and irrigated sage plains make very fertile soils; the disintegrated felspathic and limestone make a rich, porous, and absorbent earth, if well watered. The Mormons now manufacture almost everything they use, even to articles of silk; the precious metals, coal, iron, and building stones are abundant, and the water-power for machinery is ample.

The Tabernacle will hold about 10,000 persons; it is the first object seen when approaching the city—its bell-shaped top looking like a balloon rising above the trees; the building is oval, 250 by 150 feet, the roof supported by forty-six columns of sandstone, from which it springs in one unbroken arch, said to be the largest self-sustaining roof on the continent; the height of the inside is 65 feet. It contains an organ, second in size only to the Boston organ, made by a Mormon in Salt Lake City. The seats are plain, those of the men and women separate. The foundations of the great temple are laid in granite, and are now even with the ground, above which it is doubtful if they rise; the building was to cover about half an acre, and to be one of the grandest church edifices in the country; the main structure 100 feet high, with three towers on each end, the central one 200 feet high. The fine granite of which it was to be built resembles the Quincy sienite, but is much whiter; it is found in abundance in the neighboring mountains. The theatre, city hall, and council house, are fine structures, and many of the stores compare favorably, both inside and out, with our own.

Though Capt. Stansbury, in 1850, mentions seeing myriads of wild geese, ducks, and swans on the surface of the lake, I saw nothing but a few ducks and snipes around the edges, scarcely disturbed by the noise of the train. The shore is naked and bleak, and there are none of the invigorating breezes of the ocean coming from its vast and motionless expanse. Except the valleys at the southern end of the lake, the country seems very barren, without fresh water, and so little elevated above the lake that a rise of a few feet in its waters would flood an immense extent of country — the only use of which would seem to be, in the language of Capt. Stansbury, that, from its extent and level surface, it is good for measuring a degree of the meridian. The lake is said to be rising annually, and the Salt Lake problem may ere long be solved by geological agencies, the people being actually drowned out.

The existence of a salt lake in this region has been known for nearly two centuries. The water is so salt, that twelve hours' immersion will so far corn beef that it can be kept without further care, even when constantly exposed to the sun; in a few days it may be made perfect "salt junk"; if the meat were only there, a "Salt Lake Meat Preserving Company" might profitably be established near these waters. There is no life in the lake, and but little in the surrounding brackish waters, so that pelicans and gulls which breed on the islands must go at least twenty miles for food for themselves and young. The water, from its density, is very buoyant, as in the Dead Sea; it is easy to float in it, but hard to swim, from the tendency of the legs to come up and the head to go down; the brine irritates the eyes, and almost chokes you if accidentally swallowed; the most expert swimmer would soon perish in its heavy waves. It contains more than twenty per cent. of pure salt, with very little impurities; if the people are not the "salt of the earth," the water is, and probably ere many years this region will be the seat and the source of a profitable and extensive industry from its natural salt works.

After leaving Ogden, and pursuing your way westward on the Central Pacific Railroad, you pass through a well-cultivated Mormon country, getting fine views of the lake, near which the track passes for miles. In nine miles you arrive at Corinne, a lively gentile town, the centre of valuable mining interests in the neighboring territory of Montana on the north. After crossing Blue Creek on a trestle bridge 300 feet long, over many sharp curves and through deep cuts, you come close to the graded bed of the old Central road, which ended at Ogden and is now unused. Here you begin to rise till you get to Promontory Point, one of the most difficult passes on the road, and near where the trains from the east and the west met May 10, 1869, when the last tie was laid which bound the Atlantic to the Pacific. This was certainly one of the most remarkable events in the history of travel; we all remember how the country rejoiced, some cities quietly and economically, like Boston, others noisily, and with generous and hospitable exultation, like New York and Philadelphia, when the message flashed over the wires on that day that the last spike was driven; the President of the road stood there in the wilderness holding in his hand the silver hammer to whose handle was attached the telegraph wire, and when he struck the golden spike at noon, the joyful news went on lightning wings to every city of the land; the locomotives screamed and rubbed their sooty noses together, and the crowd huzzaed, shook hands, drank toasts, and exhibited the hilarious and almost frantic transports peculiar to such occasions outside of staid New England. This point is fifty-three miles from Ogden, 1,084 from Omaha, and 2,730 from Boston.

At 100 miles you are about in the middle of the "Great American Desert," where the eye searches in vain for signs of animal or vegetable life; alkaline beds, sandy wastes, and rocky hills, constitute the landscape; this desert was once evidently the bed of a great salt lake, and such as would be presented were the present Utah and Great Salt Lakes to be drained, and raised to the same level.

In 150 miles you leave Utah, and enter Nevada Territory, and at Toano, 183 miles, you enter the Humboldt division of the road, ascending the desert by the Cedar Pass to Humboldt Valley, at Pequop, being on the third high point, 6,210 feet above the sea. From this there is a gradual descent, along which you obtain fine distant views of the beautiful valleys in the range, well supplied with lakes, and famous for their fine crops. The celebrated Humboldt Wells are 218 miles from Ogden; here the emigrant trains used to stop after the hard journey across the desert; there are about twenty wells, in a charming valley, in which the water rises to the surface, slightly brackish; they are exceedingly deep, and are evidently craters of extinct volcanoes, whose existence is proved by the broken masses of lava and granite all around. This valley, which seems like Eden after crossing the dry and dreary desert, is named from the Humboldt River, which, rising in the neighboring mountains, runs through it; the track follows the river for many miles. At Elko, 275 miles, stages may be taken for the famous White Pine District; Treasure City, 125 miles to the south, is the centre of extensive gold and silver mining. At Humboldt Cañon, or the Palisades, about 300 miles, the scenery is fine, much like that of the Echo and Weber Cañons on the Union Pacific Road, but more dismal from the greater bleakness and bareness; it is gloomy and grand, from the furious river which rushes along in the deep gorges. A peculiarity of the rivers here is that they spread into shallow lakes, and in summer disappear in what are called "sinks"; probably most of their water escapes by the great evaporation, though there may be in some cases a sinking into a subterranean channel, or into the absorbent sand.

As the Truckee region is approached, fine growths of timber begin to appear, clothing the slopes of the Sierra Nevada range, which you now begin to ascend; the river is extremely pretty in its rocky bed, though much of the beauty of the scenery is lost, unless the moon be shining, by passage in the night and early morning. At Reno, 590 miles, you may take the stages for Virginia City and Gold Hill, Nevada, where are the famous Ophir and Comstock silver mines. Soon after passing Verdi, following along the numerous curves of the river, and crossing several picturesque bridges, at 610 miles, you enter California. You are now ascending all the time, amid grand scenery, with mountains on each side, timber-clothed ravines, and here and there a strip of meadow. At Truckee, 623 miles from Ogden, and 120 from Sacramento, you are 5,900 feet above the sea; this is the centre of a great trade in lumber, as the best of material is abundant and accessible, and the water-power ample. Here you may start for Lake Tahoe, a beautifully clear sheet of water, very deep (in some places 1,700 feet), twenty-two miles by ten; it is part in Nevada, and part in California; this is the lake which Mark Twain so extols above the Italian lakes in the "Innocents Abroad," to which admirable burlesque the reader is referred for fuller description. Donner Lake, smaller, but as beautiful, and seen from the track, has a melancholy interest, from the domestic tragedy connected with it; here, in the early times of immigration, a party from Illinois were hemmed in by the snow; most escaped, leaving a Mr. Donner, his wife, and a German ; when a party reached the place the following spring, Mr. Donner had died, and the German is said to have been found eating a part of Mrs. Donner's body, whom it is believed he murdered. Both these lakes are probably in craters of old volcanoes, closed by some geological convulsion which has occurred in the Sierra.

The summit of the range is fourteen miles distant, and the doubling of the locomotives shows that work is to be done ; up you go constantly, getting glimpes of the lake and the mountains, till you get to the provoking snow-sheds, which for forty miles protect the road from avalanches of snow, but not of hard words from travellers, who are by them deprived of the magnificent views. You cross the range at Summit, 7,242 ft. high, 1,700 miles from Omaha, and 105 from Sacramento. The peaks of the Sierra are far above the level of the Donner Pass, and are here and there covered with snow. The Summit Tunnel, the longest of several, is 1,700 feet, nearly one-third of a mile ; the forty miles of snow-sheds, of solid timber, are said to have cost $10,000 a mile. You are now descending all the time, sometimes quite abruptly. Just after leaving Alta, sixty-two miles from Sacramento, you enter the "Great American Cañon," one of the grandest in the Sierra, where the rocks, 2,000 feet high, give a narrow passage to a branch of Feather River ; the scenery is very fine, and there are no sheds to intercept the view. Here you come to a succession of strange names, suggestive of the lively times of twenty years ago,—such as Dutch Flat, Little York, You Bet, Red Dog, Gold Run, Cape Horn. This is the region of hydraulic mining, and you see ditches and flumes, with rapid streams from the mountains running for miles to various claims, and then directed through discharge pipes with great force against the gold-containing bank, washing away immense amounts of dirt into the long channels, where the gold gradually settles from its greater weight. Chinese miners and their cabins frequently meet the eye. Going rapidly down, almost on the edge of a precipice 2,500 feet deep, you come to and double Cape Horn, the road cut into the very side of the mountain by the Chinese ; it makes one shudder to think of the consequences of the train getting off the track as it rushes with frequent screams down the steep and narrow line, around the sharp curves, and over the apparently delicate bridges ; if quicker, it is perhaps more dangerous than doubling the point of South America. Let us hope that familiarity will not breed contempt of danger, for inevitable destruction would be the result of an accident here.

The fine fruit, bottles of wine, grapes, and grain fields show that we are in one of the great valleys of California. We soon rush into Sacramento, only fifty-six feet above the sea, having descended over seven thousand feet in one hundred miles. Sacramento is the heart of California, depending on its never-failing agricultural and mineral resources ; while San Francisco is rather a great commercial market, constantly fluctuating, and as much injured by the Pacific Railroad as Sacramento, the capital, has been increased by it. It has suffered greatly from floods, from the filling up of the river by the results of mining operations; but it is now raised fifteen feet above the highest level of the river, and is now considered safe from floods. Thence to San Francisco, *via* Stockton, over the Western Pacific Railroad, is 138 miles; thus, the distance from Boston to San Francisco, nearly 3,600 miles, may be passed over, if necessary, in seven days.

The Pacific Road was in running order seven years before the limit of the construction time, the track having been laid, and well laid, at a rate before unparalleled. In twenty-two hours, on the Union Pacific Road, seven and a third miles were laid; and on the last day but one, May 8, 1869, the Chinese laid, on the Central Pacific road, ten miles of track in twelve hours. When we remember that the great road from Vienna to Trieste, over the Soemmering Pass, less than three hundred miles, and with an elevation of only 4,400 feet, required fifteen years for its construction by the Austrian Government, with all the advantages of a populous country, and then consider that our road, more than six times as long, rising nearly twice as high, and built through a waterless, woodless desert, infested by hostile Indians, by private enterprise was completed in seven years, it is truly marvellous, and a convincing proof of the wonderful energy and foresight of the American people. The completion of this road not only unites the Atlantic and Pacific, changing the course of commerce from the East Indies, but opens vast resources of our country's agricultural and mineral wealth, and brings within the reach of travellers and invalids the magnificent scenery and bracing air of the Rocky Mountains and Sierra Nevada — leading to the great natural wonders of the parks of Colorado, the Salt Lake Valley, the Yosemite Valley, with its waterfalls and stupendous heights, the giant trees, the splendid Pacific shores, the beauty of the coast ranges, and the marvels of the Columbia River and the Cascade Mountains.

## YOSEMITE — HISTORICAL SKETCH.

BEFORE describing the Yosemite Valley, it may be of interest to the reader to know something more of the history of the discovery of this wonderful locality, within a few years known only to the Indian tribes. The following historical sketch is condensed from the "Geological Survey of California," published by authority of the Legislature.

In the year 1864, Congress, influenced by intelligent citizens of California, passed the following Act:

" *Be it enacted by the Senate and House of Representatives of the United States of America, in Congress assembled,* That there shall be, and is hereby, granted to the State of California, the ' Cleft ' or ' Gorge ' in the Granite Peak of the Sierra Nevada Mountain, situated in the County of Mariposa, in the State aforesaid, and the head waters of the Merced River, and known as the Yosemite Valley, with its branches and spurs, in estimated length fifteen miles, and in average width one mile back from the main edge of the precipice, on each side of the Valley, with the stipulation, nevertheless, that the said State shall accept this grant upon the express conditions that the premises shall be held for public use, resort, and recreation; shall be inalienable for all time; but leases not exceeding ten years may be granted for portions of said premises. All incomes derived from leases of privileges to be expended in the preservation and improvement of the property, or the roads leading thereto; the boundaries to be established at the cost of said State by the United States Surveyor-General of California, whose official plat, when affirmed by the Commissioner of the General Land Office, shall constitute the evidence of the locus, extent, and limits of the said Cleft or Gorge; the premises to be managed by the Governor of the State, with eight other Commissioners, to be appointed by the Executive of California, and who shall receive no compensation for their services.

"SECT. 2. *And be it further enacted,* That there shall likewise be, and there is hereby granted to the said State of California, the tracts embracing what is known as the ' Mariposa Big Tree Grove,' not to exceed the area of four sections, and to be taken in legal subdivisions of one-quarter section each, with the like stipulations as expressed in the first section of this Act as to the State's acceptance, with like conditions as in the first section of this Act as to inalienability, yet with the same lease privileges; the income to be expended in the preservation, improvement, and protection of the property, the premises to be managed by legal subdivisions as aforesaid; and the official plat of the United States Surveyor-General, when affirmed by the Commissioner of the General Land Office, to be the evidence of the locus of the said Mariposa Big Tree Grove."

This Act was approved by the President, June 30, 1864; and soon after, Governor Low, of California, issued a proclamation, taking possession of the tracts thus granted in behalf of the State, appointing commissioners to manage them, and warning all persons against trespassing or settling there without authority, and forbidding the cutting of timber, and other injurious acts. The necessary surveys were made, and the limits of the Valley and the Mariposa Grove were established in the same year.

The grant by Congress had no validity until the State, by its Legislature, had solemnly promised to accept the trust, forever binding when once accepted.

At the next session of the California Legislature, an Act was passed accepting the Valley and the Grove, on the conditions imposed by Congress, and containing provisions for the punishment of persons committing depredations on the premises, and appointing a guardian of the grant. Since the passage of this act, the vandalism of those who would have destroyed the grove, who would have cut down a giant tree to build their houses, has been in a great measure arrested; visitors, however, may remember a huge pine prostrate near the upper hotel in the Valley, cut down in the winter of 1869–70 by persons whom Mr. Galen Clark, the guardian, had succeeded in placing in the hands of justice.

The whites living on the streams near the Valley, as early as 1850, had been greatly harassed by the scattered Indians in this region, and finally formed a military company to expel them from the country. As the Indians were pursued it became evident that they had a safe retreat high up in the mountains, and it was determined to trace them to their refuge; this was found to be the Yosemite Valley, which thus came to be known to the whites. In the spring of 1851 an expedition, under the command of Captain Boling, started to explore this Valley and to drive the Indians out of it; guided by an old chief, Tenaya, whose name is given to one of the cañons of the Merced River, they reached the valley, and drove the Indians from their supposed impregnable retreat, killing a few, and making a peace with the rest—this, it will be seen, was fourteen years before the Act of Congress, above referred to. The Indians again becoming troublesome to the miners, another expedition was fitted out for the Valley in 1852, by the Mariposa Battalion; some of the Indians were killed, and the rest fled to the Mono tribe, on the eastern side of the Sierra; having stolen some horses from their friends, the Monos pursued them back to the Valley, where a bloody battle was fought, resulting in the almost entire extermination of the Yosemite tribe.

According to Dr. Bunnell, the Indians in and around the Valley were a mixed race, made up by refugees from many widely-scattered tribes; each family is said to have had a tract set apart for its use, which had its own name; all we know of their language is preserved in the sonorous and often musical names given to the waterfalls and rocks, as elsewhere stated, which, however, have in most cases been replaced by Spanish and English names.

The visit of the soldiers did very little toward opening the Valley

to public notice; their wonderful stories found their way into the
newspapers, but were passed over as the exaggerations so often published by travellers in distant regions, where there is no liability of con-
tradiction by eye-witnesses. Mr. J. M. Hutchings, who has been long
identified with the history of the Valley, and who now keeps a hotel
there, seems to have been the first, in 1855, to collect a party of
tourists to visit the Yosemite for pleasure; in the same year, another, and a larger, party from Mariposa went into the Valley. In
1856, the regular pleasure travel may be said to have commenced —
if it can be called pleasure to toil up and down steep ridges, dangerous on horseback, at that time, and very fatiguing on foot. The
trail from Clark and Moore's hotel is even now abominable, and un-
necessarily so; fallen trees might be removed, rolling stones picked
out, fords levelled, mud holes made safe, and projecting rocks
knocked off, at very little cost of time or money. It seems unbe-
coming in the State to allow such neglect of the trails, now that the
visitors number thousands, and many of them ladies, in the course
of the summer. Mercy for the horse, as well as for the rider, de-
mands more care to be devoted to these trails, which seem now as if
purposely made to wrench, torture, and fatigue the poor traveller,
and compel him to stop at the houses of entertainment along their
course. Were the trails properly attended to, it would be easy
enough to go from Clark's into the Valley in a day; now it is very
hard to do this, and by the time they have gone twelve miles, most
travellers are weary enough to rest at the "Half-way House," and to
make the other twelve miles on the next day; like a Chicago train,
which generally contrives to get you in an hour too late to make
your Eastern or Western connection, thus compelling an unnecessary
expenditure there, this trail seems to be neglected intentionally for a
similar end.

The first house built in the Valley, in the autumn of 1856, opposite the Yosemite Fall, is still standing, and is occupied as a hotel.
In 1860, Mr. J. C. Lamon took up his residence in the Valley, where
he now lives, a lonely bachelor, in a comfortable log house. He has
truly made the wilderness to "blossom like the rose," and has succeeded in raising excellent vegetables, and some exceedingly fine
berries, and other fruit; his garden is one of the "sights" in the
Valley, and the visitor is always sure of a welcome reception; if the
proprietor be not at home to sell you his fruit, you are allowed to
pick and eat, but not to carry away, in his garden, depositing on his
window a quarter or half-dollar in silver. He thinks that he has a
claim to the tract cultivated by himself, and considers himself a *bona
fide* settler; of course he has no legal claim, as the land was not
open to pre-emption, never having been surveyed and put into the
market. Many summer residents have since put in their claims,
which are invalid under the United States laws, for the above reason,
and also because they were not accompanied by permanent residence.
None of the claimants, it is hoped, will be allowed to have their pre-
tensions recognized by Congress, or in any way sanctioned by public
opinion. The gift of Congress is too precious to the State and to the

country to be hampered by the restrictions which would inevitably be imposed by the greed of individual owners or lessees, who would surely manage it for private benefit, and not for public good. In the language of the "Survey," "As the tide of travel in the direction of this wonderful and unique locality increases, so will the vexations, restraints, and annoying charges, which are so universal at all places of great resort, be multiplied. The screws will be put on just as fast as the public can be educated into bearing the pressure. Instead of having every convenience for circulation in and about the Valley — free trails, roads, and bridges, with every facility offered for the enjoyment of Nature in the greatest of her works, unrestrained except by the requirements of decency and order — the public will find, if the ownership of the Valley passes into private hands, that opportunity will be taken to levy toll at every point of view, on every trail, on every bridge, and at every turning, while there will be no inducement to do anything for the public accommodation, except that which may be made immediately available as a new means of raising a tax on the unfortunate traveller. . . . The Yosemite Valley is an exceptional creation, and, as such, has been exceptionally provided for jointly by the Nation and the State ; it has been made a National public park, and placed under the charge of the State of California. Let Californians beware how they make the name of their State a by-word and reproach for all time, by trying to throw off and repudiate a noble task which they undertook to perform — that of holding the Yosemite Valley as a place of public use, resort, and recreation, inalienable for all time ! "

A few years since, some scientific men, familiar with California, and especially with this Valley, undertook to obtain the signatures of their fellows throughout the land, and of those connected with learned societies, remonstrating against the enormity of permitting the claims of private individuals to stand in the way of the reservation of this Valley as a public park forever. They were successful in obtaining the approval of the great majority of American savants, scholars, and eminent men ; and it is to be hoped that Congress will never recognize such claims. It would be better far to pay ten times their estimate of alleged improvements, and to secure the right of the nation to the full control of every portion of the Valley and its surroundings mentioned in the Act of Congress of 1864.

## YOSEMITE VALLEY.

THIS unique and wonderful locality, visited by the writer in July, 1870, was once the stronghold of the Yosemite tribe of Indians, who were expelled from it in 1851, and exterminated in 1852, by the whites, exasperated by their murderous attacks, and by the rival tribe of Monos. Before this time it was unknown to the whites. A few of these Monos now live in the valley, belonging to the so-called diggers, a miserable, drunken, and fast-disappearing race, living chiefly upon fish from the Merced River, acorns, and the seeds of a species of pine, called the nut-pine.

The word Yosemite, meaning a large grizzly bear, was probably the name of a chief, who gave his name to the tribe, and the valley is now called by the Indians Ahwahnee, and not Yosemite; and even the latter is sometimes pronounced Yohemite by the Mexicans. It was first visited for curiosity or pleasure in 1855, since which time the number of visitors has annually increased, so that three hotels are now hardly able to accommodate them. It is a toilsome, fatiguing, and, in many respects, a very disagreeable journey, but when carriage-roads are extended, railroads built, and the trails made decent for horse and man, it may be undertaken by the most delicate and timid with safety and delight. It belongs to the State of California, granted by Congress, and accepted by the Legislature of the State, in 1864. There are some who lay claim to a considerable part of the best portion of the valley; and should they succeed in establishing their claims, the fleecing system of Niagara would be likely to prevail, and a price have to be paid for every trail, bridge, and advantageous point of observation. It should be under the sole control and management of the State; and the sooner the State takes the roads and trails in hand, the better for its own credit and the comfort of travellers.

On account of the chilly winds rushing in from the northwest through the " Golden Gate " to supply the place of the heated current, which ascends along the coast range, the summer (July and August) is the coldest, dampest, foggiest, and most disagreeable part of the year in San Francisco; so that, going eastward, you rise several thousand feet in an air actually warmer than on the coast, and on the highest part of the Yosemite range, 7,400 feet, it is even warm in midday in summer. At Clark's Hotel, outside the valley, and at the hotels in the valley (each about 4,000 feet high), the thermometer indicated 80 deg. for six hours every day, though the nights were cool, but indescribably clear and exhilarating. At this season the traveller is sure of good weather, as rain is extremely rare, and clouds uncommon. One is impressed with the subtropical character of the vegetation on the Pacific in latitudes where, on the Atlantic, the flora of the temperate zone prevails; in Stockton, figs grow luxuriantly in the open air, and in one of the squares was a magnificent American aloe, at least forty feet high, whose beautiful yellow flowers were the pride of the city; this in latitude 38 deg. In San Francisco, in about the same latitude, the climate is cooler; Stockton is on the east side of the coast range, in the San Joaquin Valley, but of about the same elevation, as well as latitude, as San Francisco.

Among the health inducements for travel here are the invigorating air, the pure cold water, and the exercise, which, though often severe, cannot fail to strengthen an ordinary traveller, refreshed as he is, at night, by excellent food and comfortable bed; when to these is added the grand and beautiful scenery in this immense panorama of mountains, surely no further inducement is necessary for one to journey to this valley, brought within a week's easy travel of the farthest Atlantic seaport. In the words of Prof. Whitney, "Nothing so refines the ideas, purifies the heart, and exalts the imagination of the dweller on the plains, as an occasional visit to the mountains. It is not good to dwell always among them, for 'familiarity breeds contempt.' The greatest peoples have not been those who lived on the mountains, but near them. One must carry something of culture to them, to receive all the benefits they can bestow in return. As a means of mental development, there is nothing which will compare with the study of Nature as manifested in her mountain handiwork." Beside the grandeur of the mountains, and the stateliness of the trees, the most beautiful feature is the system of waterfalls, fed by the snow, which is seen glistening on the higher summits in midsummer; as the snow gradually lessens with the advancing summer, the volume of water diminishes, and, by July, some of the most beautiful, like the "Virgin's Tears," and the falls of the "Royal Arches," and the "Sentinel Peak," are entirely dried up, and even the great Yosemite, the Bridal Veil, the Vernal, and the Nevada Falls, are comparatively small by the month of August. The fact is simply alluded to here, as, in another place, more space will be devoted to this topic.

The mountains, which look so massive and uniform in outline in the distance, when approached, are found to be deeply cleft by valleys and narrow cañons.

This whole mountain system, called by Prof. Whitney the "Cordilleras," is between the Pacific Ocean and 105 deg. west longitude, including the Rocky Mountains proper on the east, and, as we proceed westerly, the Sierra Nevada and the broken region between, and the most westerly coast range.

Beginning on the Pacific, the coast ranges are geologically newer, according to the California geologists, than the Sierra Nevada, and have been subjected to great disturbances up to a comparatively recent geological period; there are in them no rocks older than the cretaceous, this and the tertiary making up nearly their whole body, with some masses of volcanic and granitic material, neither forming anything like a nucleus, or core. They have no lofty peaks in Central California, Mt. Hamilton, near San Jose, being only 4,400 feet, and Monte Diablo, so conspicuous from San Francisco, only 3,860. The scenery is picturesque, but not grand, and especially remarkable for the beautiful valleys, or parks, between the ridges, with magnificent

forests of oaks and pines; the ridges being bare. North and south of
the central region, the elevation is greater, even to eight thousand feet,
but yet not within six thousand feet of Mt. Shasta, of the Sierra Range.
The phenomena of erosion are well marked, it is said, and the atmosphere has the indescribable exhilarating property which so delights the
traveller and strengthens the invalid.

The Sierra Nevada, or the snowy range, forms the western edge of
the great continental upheaval, or plateau, on which the "Cordilleras"
(as just explained) are built up; the Rocky Mountains form the eastern edge of the same plateau, the width between the two, traversed
by the Pacific Railroad, being about one thousand miles. In this
range the peaks are the highest, and the subordinate ranges the most
regular. The base of the Rocky Mountains is four thousand feet
above the sea level, with such a gentle ascent from the Missouri River
that you hardly perceive it as you speed along for six hundred miles;
but on the west side of the Sierra you descend very rapidly, and, in
many places, apparently dangerously, seven thousand feet in less than
a hundred miles to the level of the sea. The Sierra Nevada strictly
belongs to California, being called the Cascade Range to the north in
Oregon and Washington Territories, and to the south losing itself,
more or less, in the coast ranges; from the Tejou Pass to Mt. Shasta
is 550 miles, the last one hundred being the Cascade Range — the
average width of the chain is eighty miles, taking in the lakes on the
east and the foot-hills on the west. The western slope, in the centre
of the State, rises one hundred feet in a mile, or seven thousand feet
in a horizontal distance of seventy miles; in the southern passes the
slope is much steeper than this. Donner Lake Pass, where the Central Pacific Railroad crosses the range, is about seven thousand feet
above the sea; the crest of the range is five hundred to a thousand
feet higher than the passes, or eight thousand feet high. The central
mass is chiefly granitic, flanked by metamorphic slates, and capped,
especially to the north, by volcanic materials; the activity of the subterranean forces is now indicated by occasional severe earthquakes,
more severe and more dreaded than we in the east dream of, by hot
springs and geysers, and by the existence of many well-formed, but
extinct, craters.

The scenery of the "High Sierra," as you stand upon the "Sentinel
Dome," or "Glacier Point," is very different from that of the higher
Alps. You see much less snow and ice, and no glaciers extending
into the valleys. But the rocks, even to the edge of the Yosemite, are
grooved and polished, showing the former existence of an immense
sheet of ice. You see no grassy slopes between the forest and the
snow, but the woods extend much higher up, and abruptly terminate
with the bare rock in summer, and the snow line in winter; the trees
are large, but sombre and monotonous, growing even at a height of
7,000 feet. Though there are many beautiful valleys along the
streams, and magnificent waterfalls, the character of the scenery is
rather grand, sublime, and awful, than beautiful or diversified; the
heights are bewildering, the distances overpowering, the stillness
oppressive, and the utter barrenness and desolation indescribable.

One of the most striking features of the scenery on the edge of the valley, is the concentric structure of the granite in the so-called "Domes," and "Royal Arches," of which more hereafter. Suffice it to say here, that the rounded, dome-shaped masses contrast remarkably with the sharp peaks above and beyond them; they rise from three to five thousand feet above the valley, presenting toward it a sheer precipice of nearly this height — domes of the most graceful curves, and on a stupendous scale. This concentric structure, according to Whitney, is not the result of the original stratification of the rock, and there are no evidences of anticlinal or synclinal axes or marks of irregular folding; the curves, arranged strictly with reference to the surface of the masses of rock, show, according to him, that they were produced by the contraction of the material while cooling or solidifying, giving one the impression that he sees the original shape of the surface. The concentric granite plates overlap each other, absolutely preventing ascent from the valley; as these immense plates have fallen, some from a height of over 3,000 feet, detached by the frost, and other agencies, they have left the enormous cavities which have received the name of the "Royal Arches," and royal indeed they are.

All observers agree that the snow disappears from the highest summits rather by evaporation than by melting, and that the air there is remarkably dry; and by this is explained the general absence of glaciers on Mt. Shasta and similar elevations, where in the Alps glaciers would exist; immense masses of snow, miles long and hundreds of feet thick, remain all summer, thawing and freezing on the surface, gradually wasting away without becoming glacier ice, and yielding comparatively small streams of water. Still, at a comparatively recent geological period, immense glaciers existed in these mountains, and the usual traces of scratched and polished surfaces are common enough, and moraines of great extent are found — these evidences of former glacial action, however, seem to be limited to the higher parts of the range, and not to descend below 6,000 feet above the sea, except in a few exceptional cases, where the configuration of the upper valleys was favorable to the accumulation of large masses of snow — this indicates at that period a considerably moister climate than now exists there. Glaciers extended from Mt. Dana (13,000 feet above the sea) to a level of the upper border of the great Yosemite, or 7,000 feet above the sea, the bottom of the valley being 3,000 feet lower. The weight of an ice sheet a mile in thickness, may have had something to do with the sudden subsidence which many geologists think was in part the cause of the formation of this valley. Marks of glacial action are manifest on the "Sentinel Dome," and on "Glacier Point," both groovings and polishings; this polishing extends far down the smooth surface on the south side of the valley, near the Illilouette Cañon, a steep, gigantic slide for a thousand feet, of perfectly smooth rock, which makes one dizzy to look at from above or below, ending, as it does, in a vertical wall toward the valley. There are no signs, that I know of, of glacial action in the valley. The Little Yosemite Valley, 2,000 feet higher than the big Yosemite, but greatly resembling it, communicates with the latter by the Nevada Fall, the main stream of the Merced River running through both. No doubt a glacier passed down the Illilouette Cañon from the Mt. Starr King group to the edge of the valley; the land at the head of the Merced River was not high enough for the formation of a glacier into the Yosemite Valley, and there is no evidence that it came beyond the edge, as above stated, though it doubtless filled the higher Little Yosemite.

The famous valley is about 155 miles from San Francisco, a little south of east, or 250 by the usual line of travel. It is best to stop, when coming from the east, at Stockton, distant ninety miles from San Francisco by rail. I went by the Mariposa route, the longest, with the most horseback-riding, but leading near the Mariposa grove of big trees, and affording, on the whole, the grandest views. We took a private conveyance, three of us and a driver, at Stockton, the usual charge for which is $16.00 a day, including the food and all expenses of driver and two horses; the stages are crowded and uncomfortable, (though, from experience, I think not more so than the private carriage,) but are considerably cheaper and quicker, as they travel day and night. By this route you have about twenty-five miles to go on horseback, mostly up and down steep and rough trails, to reach the valley — this we did in one day; but it is better to take two, as both horses and riders get greatly fatigued.

You cannot enter the valley without rising about 3,500 feet above the point you wish to reach, viz.: the bottom of the valley— this is 4,000 feet above the sea, and so is the ranch of Mr. Clark, from which you start; from this you ascend to 7,400 feet, and then descend about 3,500 into the valley. This severe, but necessary, toil, is what, with the dust and heat, makes the journey so fatiguing. You can do it all on horseback, as Mark Twain's pilgrims did in the Holy Land; but pity for the horse, and comfort, if not safety for the rider, impels you often to dismount, exchanging the fatigue of climbing for the weariness and soreness of the saddle (it is, for the first few days, a sort of drawn battle between the abductor muscles of the thighs in riding, and the muscles of the calves in ascending or descending on foot). The cañon of the Merced River, whose shallow and placid stream runs through the valley, has such steep sides, that a trail there is next to impossible for any one but an Indian or an Alpine climber; and so the valley has to be entered from the side, at the western extremity, either by the Mariposa trail on the south, or the Coulterville trail on the north.

## Yosemite Valley.

THE distance from Stockton to Mariposa is about ninety miles, and from there to Clark's about twenty-five, or 115 miles by stage or carriage, and then twenty-five more on horseback to the hotels in the valley — or 140 miles, carpet-bagging from your base at Stockton, which, last year, was the nearest point by rail; though probably even Mariposa will ere long be reached by rail, and a carriage-road be made twelve miles beyond Clark's, reducing the terrible horseback ride to twelve or thirteen miles. Rough as it is, many ladies accomplish it every year. A railroad has now been finished from Stockton to Copperopolis, reducing the stage ride by Coulterville about twenty-eight miles.

We left Stockton in a light carryall, with two horses, at six o'clock in the morning, intending to take our own time for the journey. On getting into the country, everything looked burnt (this was in the last half of July); the clayey soil was cracked in all directions by the heat, sometimes to a foot in depth, presenting very much the appearance of the geological mud-cracks so frequently seen in the rocks filled with a harder material. The crops were all stacked in the fields, immense piles, no barns being necessary to protect the grain at this dry season, and there they remain till the steam-thresher comes along, and the threshed grain is placed in sacks, loaded into wagons, and transported to the river or the cars. The scarcity of water at the surface gives an indescribable parched appearance to the landscape; yet there seems to be an ample supply at a moderate depth, and every farm has its wind-pump, raising water from a kind of Artesian well, distributed by gutters over the fields and gardens. The interminable barren plain, dotted with herds of cattle and horses driven by their herdsmen, the long trains of grain-laden, creaking wagons, drawn by mules, with the numerous wind-pumps lazily and noisily working, remind one of the Spanish landscape, and it would have been entirely in keeping with the surroundings to have seen Don Quixote and Sancho Panza ride forth from a court-yard. The squirrels ran out from their burrows by the sides of the road, and scampered across the fields, and occasionally a long-eared, diminutive, half-starved-looking hare would be seen picking up a scanty meal among the stubble. As we got into the country, or rather desert, for it was a hot, treeless, sandy plain, the squirrels became more numerous, apparently in inverse proportion to the amount of visible food, accompanied by the grave-looking burrowing owls which inflict their presence, the other side of the Rocky Mountains, on the prairie dogs; horned lizards were not uncommon, lively and plump, but what they found to eat I could not discover, as insect life seemed to me decidedly scanty; they may find ants, as now and then their hills were to be seen. These plains are remarkable for their mirage, and it is impossible at first to believe that the lake in advance, with its grateful shade of trees, is nothing but deception and reflection from the sand, with here and there a scraggy tree. You meet no travellers on foot except a few Chinese, dressed like ourselves, except the hat and blouse, going to and from the mining locations; and even they frequently exchange money for time, and ride by stage. Wherever a clump of trees appears, the woodpeckers and magpies are numerous, and the wild pigeons are hardly wilder than the pigeons in our streets. The oaks are beautifully festooned with a long, hanging moss, giving the same funereal look that a similar appendage does to the cypress swamps of the Southern States; unlike the latter in most respects, it also prefers dry and sandy plains instead of moist places, and is confined, as far as I saw, to the oaks. The Stanislaus and Tuolumne Rivers are crossed by ferries, moved by most primitive hand-power; "pay or stay" is the word there, and a ferry-man is even more imperturbable than the keeper of a turnpike; if travellers were numerous, the delay and the changes would be a great nuisance, and the only way to get over the difficulty would seem to *bridge it.* ·

The dust and the heat were overpowering; and, much as we suffered, the horses suffered more; but if a horse gives out there are plenty of others, and in some of the corrals there were so many that the owner did not positively know how many he had. After dinner one of the horses was used up, and with a fresh one we started again, contrary to the advice of the driver, who was not sure of his way by night, and rode consequently till midnight, having lost our way as far as the path was concerned, but sure of coming out all right by keeping the pole star over our left shoulder, as you can ride anywhere on this level plain just as you can upon a prairie. We arrived at Snelling's at midnight, and, after sleep rendered unrefreshing by public snoring and foul air, with the additional discomfort of a very poor breakfast, we began the second day, equally hot and even more dusty, but more interesting as the region became hilly. At noon we had reached Hornitos, well named, as it is truly a "little oven," and gave us a good baking; passing from this through Bear Valley, you traverse the famous Mariposa Estate, where fortunes have been lost and won; the former rich gold placers have yielded up their wealth, and the region is in a state of decay, given up principally to Jews and publicans, and the Chinese; the latter patiently, and laboriously, and successfully digging over the old sites, already dug over many times before; yet with their sobriety, economy, and perseverance, picking up many a "chispa" (or sparkling bit of gold) overlooked by the more hasty American diggers. There is, I believe, only one stamp mill on this immense property, and that not doing much. Deserted huts, dilapidated flumes, broken mining apparatus, and desolate heaps of stones, speak sadly of the crowds that have departed without the treasure which they sought; in fact, the whole region, especially near the watercourses, has been dug over, and looks like a violated graveyard, fit emblem of the bright hopes there buried. The only sign of life is indicated by the turbid streams, often only a few inches deep and wide, discolored by the washings of the indefatigable Chinese, not far off.

At Mariposa, which is situated in a charming valley, though at this season very dry, hot, and dusty, we found another relic of the olden time in a double wheel of about twelve feet diameter, and two feet wide, covered with lattice-work, set up in the back-yard of the principal hotel. In this was gravely walking, as in a treadmill, a large dog, turning the wheel slowly, thus acting upon a pump which supplied the water for household purposes — somewhat in the manner of the dog-turnspit of old. The work seemed easy, and the dog was sleek, and apparently contented to perform his welcome duty for the house.

Here you start by stage or your own conveyance, for the higher hills, for White and Hatch's, twelve miles distant, 3,000 feet above the sea; after a good meal and welcome rest there, you start again for the mountain region, and very soon come among the tall pitch pines, with their grateful balsamic odor, and ascend nearly 3,000 feet more in about seven miles, and then rapidly descend in four or five more, by a good but very zigzag road, 1,700 feet to Clark and Moore's, the real starting-point for the valley and for the Mariposa Big-Tree Grove. You generally arrive here in the evening, and the coolness of the air and water are very grateful after the heat, and dust, and jolting of the day; the house is kept by New England people, and you are received in the most hospitable manner, and nothing is wanting to make you comfortable. Mr. Clark is the guardian of the grove, appointed by the State. You here, if you wish, mount your horse for the grove, about four miles distant; but of this I may speak on another occasion; here also is the south fork of the Merced River, inviting you to a bath in its clear cool water, and very few, I think, decline the invitation to get rid of the accumulated dust of the journey from Stockton. The hotel is about on the same level as the Yosemite Valley, but many a weary mile and aching muscle intervene, for here you take horse. Leaving early next morning, you cross the river, and in about four miles ascend 1,900 feet, where you cross Alder Creek, stopping to give yourself and horse a drink. You then ascend to Empire Camp, now used only as the house of the tenders of the sheep here kept; we went through one flock containing several thousand, and the dust they kicked up was suffocating, as it was quite impossible to go on without trampling upon them in the narrow path, until the flock had passed; the grizzlies must have fine pickings among them. We met, also, horses by the score, running wild, turned out to recover from the fatigue of carrying pilgrims like ourselves, and many very much heavier, up and down these terrible hills. You then, after about twelve miles, arrive at the half-way house, or Perigo's, 3,100 feet above Clark's, and 7,100 feet above the sea; here frost appears early in August, preventing the production of any useful crops, but apparently admirably suited to the chipmunks, or striped squirrels, which run in and out the sheds and houses like mice. The guides and horses are obliged to remain out of doors at night, the former consoling themselves by a large bonfire. From this you may branch off to "Sentinel Dome" and "Glacier Point," though it is better to make this trip after you have seen the valley, as it is better enjoyed after you know what you are looking at — it is like a review of a subject previously studied, the principal points of which cannot be understood or appreciated until you have personally examined the whole field of observation.

Going forward, then, you enter Westfall's Meadow, a very dangerous place out of the path, even in the dryest time of the year, from the liability of miring or even drowning your horse, and perhaps yourself — it lies in a basin between two high ridges, and is never dry. By day the wind blows up the mountains, and by night down; you have the dust, therefore, always with you going up, and also going down if any one be in advance of you; this dust is the greatest annoyance of the trip. When you have ascended 3,426 feet above Clark's, or 7,400 feet above the sea, you come suddenly to what is called "Inspiration Point," and there the magnificent panorama of the valley at once, and for the first time, bursts upon the view; no language can describe its grandeur, and no painting can do it justice; the best idea is given by the excellent photographs which have been taken from this point, but even these are poor in comparison to vision, and serve rather to recall features once seen than to depict the great reality. It is well called "Inspiration Point," for it is an inspiration even to those familiar with the grandest mountain scenery; it is probably the most magnificent view to be had in the world. Having reached this point, where the exploration of the valley really begins, what is seen in the valley will better be described on another occasion; and I will only add a few remarks, which may be interesting to those who intend or hope to visit it, comparing the advantages of the two principal routes, the Coulterville and the Mariposa. By the Coulterville route which enters the valley from the north, you have more and finer views of the distant Sierra to the north and east, and see the various points of beauty in succession; by the Mariposa trail, you go near the big trees, and the whole grandeur of the Yosemite is revealed at Inspiration Point; if you return by the Mariposa route you get a second view, or rather review, as a whole of what you have visited in detail, and, besides, can easily make the grand trip to the Sentinel Dome and Glacier Point, the view from which is nearly as grand, perhaps, as that from Inspiration Point. If one prefers to try both, enter by all means by the Coulterville, and leave by the Mariposa route. As to public conveyances, you leave Stockton at six A. M., and reach Hornitos about eight P. M.; starting next morning, you arrive at Mariposa at noon, and at Clark's at night. There, next morning, you take horses for the valley, distant twenty-five miles, and do it in one or two days, according to the tenderness of the parts of the body which rub against the saddle, and your experience as a horseman. You spend three days, at least, in the valley; then one to return to Westfall's, where the trail goes off to the Sentinel Dome, which should not be omitted — one to Clark's and the big trees — then two days by stage to Stockton again — in all eleven days.

If you go by private conveyance, it takes two days longer, with

much more expense (more than twice as much), and with no more
comfort than by the stage—in fact, delay upon the road, in the dust
of summer and heat of the day, is only a prolongation of misery,
which, at the best, is very hard to bear.  In fact, the knowledge obtained by experience, in the minds of some travellers I met,
is not worth the jolting, and jamming, and bruising, and soreness,
inevitable in this journey—in fact, one of them said that though, in
the words of Solomon, if you bray a fool in a mortar with a pestle
yet will not his folly depart from him, the tumbles and bumps and
scrapes of the Yosemite trails will take the foolishness out of a man,
and the poetry too.

But, with all its fatigue and discomforts, there is nothing in this
trip to alarm the most timid person; there is no danger to the nervous system, but great fatigue to the muscles, whether riding or
walking.  Notwithstanding these drawbacks, I think no one who has
made the trip will ever regret it, though he may not, till railroads
are extended, be inclined to repeat it—when he remembers the
grandeur of the scenery, the magnificence of the forests, the extraor-
dinary beauty of the waterfalls, and the uncommon purity of the air
and clearness of the sky in these elevated regions.

As the traveller is supposed to be left now at Inspiration Point,
gazing into the beautiful valley, it may be well to allude to the sublime views from Sentinel Dome and Glacier Point, both above and
on the edge of the valley.  The Sentinel Dome is a great rounded
smooth mass of granite, about five miles to the north-east of the
half-way house of Perigo's; there are upon it a few stunted pines,
and one remarkable one on the summit, a welcome support to cling
to during the high winds which prevail there; you may ride to the
very top; but most prefer to walk, especially in descending, so slippery is the bare rock.  Looking north-east up the Tenaya Cañon, in
which is one of the forks of the Merced River, and the beautiful
"Mirror Lake," you have on the left, in the distance, the snow-covered
Mount Hoffmann, and almost under it the "North Dome," 3,568 feet
above the valley, the upper portion of the rounded, concentric-lay-
ered, granite mass before alluded to as the "Royal Arches," inaccessible from the valley, but easily ascended by a ridge which runs to
the north; this magnificent dome is worthily supported by the Royal
Arches, by the side of which man's proudest architectural monuments
are utterly insignificant.  On the right, or south border of the cañon,
is the "Half Dome," with its stupendous vertical face of 3,000 feet
from the summit, then a steep slope of about seventy degrees of 2,700
feet more, the top being absolutely inaccessible—beyond is the
Clouds' Rest, 700 feet higher, but belonging rather to the Higher
Sierra than to the Yosemite group; on the opposite side is Mount
Watkins, named from the eminent photographer of this region, and
beyond this the distant Sierra.  The Sentinel Dome is 4,150 feet
above the valley, and the Half Dome is nearly 600 feet higher.  To
the east is seen the Nevada Fall, with Mount Broderick, or the "Cap
of Liberty," to the left of it; in the far distance the Lyell group,
and to the south-east the steep, inaccessible granite peak, named
after Starr King, belonging to the Merced group.

About half a mile north-east of the Sentinel Dome, and directly in a line with the edge of the Half Dome, is Glacier Point, overhanging the valley, and presenting a view which for beauty and grandeur is by many regarded as the finest around the valley. Both the Vernal and the Nevada Falls are in sight to the east, separated from each other about a mile, and the nearest one, the Vernal, a little more than a mile from the spectator; the point is fringed almost to the edge with Jeffrey's pine. The view of the Half Dome, only two miles distant, and directly in line, is grand in the extreme. To the north is seen the Yosemite Fall, 2,600 feet high, and to the west, limiting the vision, is the massive El Capitan, a solid block of granite, 3,000 feet high, projecting squarely into the valley, with almost vertical sides. Below you see the green of the valley contrasting beautifully with the cold gray of the bare rocks, the tall pines looking like shrubs, and a man scarcely discernible. The thread of the Merced River sometimes glistens in the sun, and the garden of Mr. Lamon forms a pleasing feature with its greenness and orderly arrangement. Travellers who fail to visit this point, in my judgment, lose one of the finest views in the whole Yosemite.

### THE CLIFFS AND FALLS OF THE YOSEMITE VALLEY.

THE Yosemite Valley, according to the California geologists, is nearly in the centre of the State north and south, and in the middle of the Sierra, which is here seventy miles wide. It is nearly level, about five miles long, one half to a mile wide, and sunk nearly a mile perpendicular below the neighboring region. It is an irregular trough, with many projecting angles not corresponding with recesses on the opposite side, an argument against its being a geological fissure. At its eastern end it branches into three cañons, the Tenaya, little Yosemite, and Illilouette, down which flow three main branches which form the Merced River in the valley; the last two with fine falls, the first with a beautiful crystal lake. At the west end it is narrow and V-shaped. The walls are almost vertical, and of great height, both absolutely and compared with the width of the valley, and are remarkable for the small amount of *debris* at their base. The most distinguishing characters are the domes and the waterfalls, any one of which in Europe would be of world-wide fame; there is nothing in the Old World to compare with either, and of the latter many, far surpassing anything in the Alps, are not noticed, as there are so many fine ones demanding the traveller's attention.

Coming in from the Mariposa trail, as you descend from Inspiration Point 3,000 feet, slowly and painfully to yourself, and with pity for the horses, you come at every turn upon views of surpassing grandeur and beauty. On the left stands the massive "El Capitan," an immense block of bare, smooth, light-colored granite, 3,300 feet high, projecting squarely into the valley, and with almost vertical sides. At first you cannot realize its stupendous bulk and height; there is no standard to judge by where everything is on so grand a scale; nothing but climbing about among them will open your eyes to the amazing heights of the cliffs and falls. Of El Capitan, Whitney says "it seems as if hewed from the mountains on purpose to stand as the type of eternal massiveness. It is doubtful if anywhere in the world is presented so squarely cut, so lofty, and so imposing a face of rock." In a recess in one corner is the "Virgin's Tears" fall, 1,000 feet high, rarely seen by travellers, as the creek which supplies it is dried up early in the season; it is superior, while it lasts, to the famous Staubbach fall in Switzerland, the admiration of Alpine tourists, and one of the finest in Europe. The Indian name of El Capitan is "Tutocanula," said to be an imitation of the cry of the cranes, which in winter used to enter the valley over this rock.

Directly opposite is the beautiful "Bridal Veil" fall, about 700 feet in perpendicular height, and 200 more of cascades as it rushes over the debris at the bottom of "Cathedral Rock," over which it pours; the creek which supplies this fall, you pass when going to "Sentinel

Dome," and the coolness of its clear water is sure to be tasted by the traveller and his horse. In the dialect of the Indians, this is "Po-hono"—a blast of wind, or the night wind, from the chilliness of the air experienced by coming under the cliff, and perhaps from the swaying of the sheet in the wind like a veil; others think Pohono was an evil spirit, whose breath was a dangerous and deadly wind. Whatever its derivation, the poetical name of the Indian is, here as in other places in the valley, much superior to the English one. As in all the falls, the amount of water varies greatly with the season, be-ing greatest in May and June; it is most beautiful later in the summer, when the volume of water is small, as it then sways more gracefully in the wind.

The "Cathedral Rocks," over which the "Bridal Veil" falls, are neither so high nor so vertical as El Capitan; though only about 2,600 feet high, they are very grand whichever way you look at them; from one point the pinnacles called the "Spires" are so squarely cut that they remind you of the towers of Notre Dame in Paris. These grand masses, amid so many grander, are hardly no-ticed by the tourist; what appear on the top like bushes are ever-greens 125 to 150 feet high, as large as those which excite your wonder in the valley.

On the opposite side is a triple group of rocks, known as the "Three Brothers," rising one behind the other, the highest being 4,200 feet above the valley. The Indian name is "Pompompasus," or "Leaping Frogs," from a fancied resemblance to three frogs with their heads turned in one direction, the highest in the rear as if in the act of leaping.

Nearly opposite the "Brothers," just in the rear of the first hotel, or Leydig's, is "Loya," or "Sentinel Rock." This is a slender peak of granite, over 3,000 feet high, the upper third standing up like an obelisk or signal tower; it is one of the grandest masses of rock in the valley. Behind it, and more than 1,000 feet higher, is the "Sen-tinel Dome," before described, not seen from the valley. From "Sentinel Rock" descends a small fall, 3,000 feet high, 400 feet higher than the Yosemite fall, but reduced in July to a mere thread, unperceived by most travellers; in early spring it is a very beautiful cascade.

The great feature in the valley to most persons is the Yosemite fall, just opposite, surpassing in height all others, here or else-where, having an equal body of water. The grandeur and beauty of this fall and its surroundings are, in a measure, familiar from ex-cellent photographs, engravings and paintings. The creek which supplies the water is fed by the melting snows of the Mt. Hoffmann group, ten miles to the northeast; of course the volume of water varies greatly, being very large in spring, but in August reduced two-thirds. When generally seen, in June and July, the stream at the fall, according to Whitney, is twenty feet wide and two feet deep. The height is 2,600 feet, half a mile; a vertical fall of 1,600 feet, swaying in the wind and broken into spray in a most beautiful manner, and falling into a deep, rocky recess; then a descent, in a series of cascades, of 600 feet; and then a final plunge of 400 feet to the bottom of the valley, falling upon a rough assemblage of rocks, then flowing off to join the Merced River, being ignominiously made to turn a saw-mill on its way. All the falls you see well from "Sentinel Dome," opposite, distant two and a half miles, and considerably above them. It is impossible to imagine anything finer than this scene under a full moon.

A mile or two above the Yosemite fall, the valley branches into three cañons, the middle one kept by the main Merced River, with the "Vernal" and "Nevada" falls, the little Yosemite Valley (a miniature copy of the greater), and the ascent to the Lyell group, where the river heads; on the left hand is the Tenaya cañon, and on the right the Illilouette. Just before these branches is the "Washington Column," ("Shokoni,") about 2,500 feet high, and the "Royal Arches," ("Tocoya," or the "Basket,") supporting, as it were, the "North Dome"; the last is about 3,700 feet high, made up of huge concentric plates of granite overlapping each other. The "Half," or "South Dome," ("Tisayac,") opposite, about 6,000 feet high, is another magnificent mass of smooth, rounded granite, looking as if the western half had been split off and swallowed in an abyss — it is truly a "wonder among wonders."

Following up the Tenaya cañon, over a very rough trail among boulders and rolling and rough stones, you come to "Mirror Lake" ("Waiya"), so called from the reflection in its still, clear water of the surrounding peaks, Mt. Watkins and others. Farther up is "Cloud's Rest," nearly 7,000 feet high, connecting with the higher Sierra, and frequently surrounded by clouds when the other peaks are clear.

Returning and going up the cañon of the main Merced River, you visit the "Vernal" and "Nevada" falls, each the body of the main river. The trail is in many places difficult, but nowhere dangerous, with ordinary care; you are almost constantly ascending, winding in and out, up and down, along the banks of the stream, which flows with great rapidity and turbulence in its rocky bed, affording some enchanting views of mountain and cascade scenery. Here we met Mr. Shapleigh, an artist from Boston, with whose fine sketches most of our California tourists are now familiar.

After about a mile's climbing, you arrive in sight of the "Vernal Fall" (*Piwyack*, white water, or shower of diamonds), about 400 feet high. The granite behind the sheet is square, and little, if any, eroded by the falling water; so that it is hard to believe that this cañon and fall have been the result of any causes now in action there; there must have been a subsidence, as most observers think was the case in the formation of the valley itself. The trail up the cañon in its upper portion, around and along the steep side of the mountain, is slippery, and wet with the spray; you can ride by a rough road to the top, but most persons prefer to walk, muddy and moist though it be. You can go no farther than the base of the cliff by the path, and you willingly stop to rest and admire the ever-changing rainbows over the water, and enjoy the refreshing coolness and shade. At this point there is a spacious cavern, formed in the concentric layers of granite peculiar to this region; this was once probably the lair of wild animals, and the still wilder Indian, as it is now said to be of the rattlesnake. The ascent is now made by perpendicular and not very strong ladders of wood, making the nervous tremble lest their feet should slip, and anxious lest they should meet a rattlesnake sunning himself on the landings along the ascent. These reptiles are numerous here, and are frequently killed by the sticks with which cautious travellers arm themselves; though we met none alive, the rattles exhibited, and the dead ones hanging to the trees, show that they are too common for comfort. At the summit the view down the cañon is indescribably grand, and the more enjoyable as a parapet of granite runs along the edge, just high enough to support you in safety almost on the very brink.

Going up the stream by a very rugged and often steep path, winding around immense boulders which have fallen from the heights on each side — the beautiful Merced River foaming along in its rocky bed, with rapids succeeding each other in endless variety, in one place shooting like silver lace-work over a smooth surface into a pool of emerald hue — crossing the main and rushing stream on a rude bridge, and some of its torrents on trunks of trees, not altogether safe because steep and slippery, you come, after a mile of hard climbing, to the "Nevada" fall ("Yowiye," slanting or twisted water). This name is given because just below the edge is a projecting shelf, which receives and throws to one side a great portion of the water; this adds much to the picturesqueness of the fall, by its unusual shape. It is the grandest in the valley, having a large body of water of extreme purity, falling about 700 feet; it is surrounded by majestic mountains, the most noted of which is the "Cap of Liberty," or "Mt. Broderick" (Mah-ta), 4,600 feet high, and almost as grand as the "Half Dome." The descent between the Nevada and the Vernal falls is about 300 feet. Returning you may look up the cañon of the Illilouette, where in early spring is a fine fall of 600 feet, rarely visited, from the difficulty of the trail.

The Yosemite Valley is nearly level, sloping very gently to the southwest, the sluggish Merced River, about seventy feet wide, flowing through it; it ends in a narrow cañon to the west. It is 4,000 feet above the sea, and contains some swampy meadows supporting alders; there are also the spruce and poplar, and in the sandy parts the pitch pine, white cedar, firs and oaks. The walls are light gray, very bright in the sun, here and there discolored by organic matters in solution in the water; most paintings give the rocks a golden haze which they do not possess.

The characteristics of this valley are, as far as I know, nowhere else in the world combined on such a large scale. These are: grand perspectives; stupendous perpendicular cliffs; vast domes; glistening ribbons of cascades coming apparently from the clouds; thundering falls like the Vernal and Nevada; frightful chasms; crystal lakes; gigantic pines; and a beautiful river. There is a painful lack of color arising from the union of cold gray granite and sombre evergreens; the valley is so narrow, and the walls so high, that the sun practically sets early in the afternoon, adding a premature dusk to the wild scenery.

In early spring, when the snow begins to melt on the mountains, innumerable waterfalls appear, most of which are dried up before travellers arrive. Some prefer the grand volume of Niagara, others the graceful height of the Yosemite; both are equally wonderful and beautiful, but no more to be compared than the sturdy oak to the clinging vine, or the vigor of man to the beauty of woman. As a rule, I should say that the female sex prefer Niagara, while males prefer Yosemite, from the natural love of their opposites. The high waterfalls of Europe are not large; the highest (Gavarnie, in the Pyrenees) is not half so high as the Yosemite, and is a mere trickling stream; the Staubbach, in Switzerland, is about as high as the "Bridal Veil" (900 feet), but has very little water; the Voring Foss, in Norway, said to be the finest in Europe, is only 850 feet, and is considered, by those who have seen both, far inferior to the California falls. Beautiful as they are in summer, these falls in winter, with their frozen spray forming domes more than 100 feet high, the drops rebounding in the sun like diamonds, must present a sight of surpassing beauty and grandeur.*

How was this grand and unique valley formed?

Nowhere is the tremendous erosive action of water more fully exhibited than in the great cañons and valleys of the Sierra Nevada; cañons 2,000 feet deep have been worn in hard lava by the long-continued action of mountain torrents, and the rocks are everywhere channelled by this cause; but these gorges do not have the vertical walls of the Yosemite, nor such perpendicular granite surfaces as "El Capitan," 3,000 feet high, meeting each other at right angles; the faces here are turned down the valley, opposite to that in which erosion by water could have acted. The "Half Dome" rises vertically 2,000 feet above the level walls of the valley, and the same distance above the action of water, even had its torrent filled the whole valley. There is no apparent source of supply for the water necessary to have produced such an erosion, even upon the wildest glacier theory; the valley is too irregular and sharp upon its sides, and the cañon of exit too narrow to admit of this explanation.

The erosive action of ice cannot be reasonably advanced as the cause; there is no evidence of ice-action in the valley, though there is plenty of it on the sides above it, and to the very edge; moreover, the work of ice, as seen in the Alps and elsewhere, is entirely unlike what is seen in the Yosemite Valley.

It cannot be regarded as a geological fissure, for the walls are on an average half a mile apart, and the same in depth; and they in no way

---

* We are informed by a traveller recently returned from the valley, that the Yosemite fall was entirely dry this year in the first week of September; travellers at this season lost, therefore, perhaps the most beautiful feature of the valley, and the most remarkable waterfall in the world.

correspond on the two sides. As it is transverse to the line of the mountain upheaval, it cannot be the result of folding.

There remains the hypothesis of the California geologists, which seems to me the true one, viz. : that during, or perhaps after, the up- heaval of the Sierra, there was a subsidence — that the bottom of the valley sunk down to an unknown depth, the support underneath hav- ing been withdrawn during the convulsion. This explains the ab- sence of debris, which has gone down to fill the abyss. The valley was undoubtedly once filled with water; the disappearance of the glaciers, the gradual dessiccation of the country, and the filling up of the abyss, have converted the lake into a valley with a river running through it; the process of filling is continually going on from the ac- tion of the elements upon the surrounding rocks.

There are other examples of similar probable subsidences, as in the little Yosemite and Hetch-Hetchy Valleys. Lake Tahoe and its valley are perhaps the result of a similar subsidence, the lake occupy- ing the cup of a sunken crater.

The following, from the *Overland Monthly*, well describes the sensations which arise on viewing the Yosemite Valley :

"Such magnificence of rocks, such stupendousness of cliffs, far outstripped conception, and staggered even perception itself. You disbelieve your own eyes. Judgment fails you. You have to recon- struct it. Comparison serves you little, for you have no adequate standard with which to compare, or by which to estimate the rock- mountains before you. They are like nothing else but themselves. Look at that tree: elsewhere you would call it lofty. It must be a hundred and fifty feet high, and yet that wall of rock behind rises straight up to twenty times its height above it. Slowly you begin to "even yourself" to the stupendous scale of the gigantic shapes around, though yet trembling and staggering under the overwhelm- ing immensity pouring in upon you from around and above. A score of cataracts in solid rock, Niagaras in stone pile upon each other and pour over each other in absolutely painful tremendousness. Solidified vastness; infinity petrified; the very buttresses of eternity over- power the sight and benumb the brain. The works of God crush out the words of man. We can only silently uncover and stand speechless, with abated breath."

# BIG TREES.

NO traveller from the East should fail to visit one or the other of the groves of "Big Trees"; the principal ones are the Calaveras and Mariposa, the property and the charge of the State of California, to be held as public parks forever. These trees are the highest and largest of the vegetable kingdom, both dimensions considered; though some of the eucalypti of Australia are 100 feet higher, and the baobab of Africa is larger in diameter, the former is of comparatively small diameter, and the latter of medium height. We are familiar here with the wood and bark, and even the cones, seeds and foliage, from a large specimen recently exhibited in the principal cities and towns, and which, it is hoped, may ere long find a permanent resting-place in Boston.

These huge trees are said to have been accidentally discovered in 1852 by a hunter employed by a mining and water company, whose story was so little believed that he was obliged to lure the workmen to see the trees, by leading them to a huge grizzly bear which he said he had killed, and was unable to bring in alone.

The wonder soon got into the papers, and was quickly known all over this country and Europe. Dr. Lindley, failing to recognize its genus, named it "Wellingtonia gigantea," after the greatest modern English military commander; it had already been called in America "Washingtonia gigantea," in compliment to our noblest military hero. Decaisne, a French botanist, discovered that it belonged to the same genus as the California redwood (*Sequoia sempervirens*), and it is now known in science as *S. gigantea*.

The genus was named in honor of Sequoyah, a Cherokee half-breed, better known as "George Guess," who lived in the last quarter of the 18th, and the first third of the present century. He dwelt in the north-east corner of Alabama, and invented for his tribe an alphabet and written language; there were in it 86 characters, each representing a syllable. It was considerably used, and a paper was printed partly in these characters. The memory of this benefactor of his people will probably soon pass away with his nation, now driven beyond the Mississippi, and rapidly becoming exterminated.

The redwood, so called from the color of its wood, is limited to the seaboard, seeming to require for its growth the salt mists from the ocean. The "Big Trees" are inland, and confined to limited ranges in the Sierra; but both are Californian, and the latter entirely so. The genus is also found fossil in the earlier "tertiary" of Greenland, as high as lat. 70 deg. N.; the study of these giants, therefore, is of great interest to the palæontologist and geologist.

The redwood is found along the coast from 36 deg. to 42 deg. N. Near San Francisco and the large towns they are all cut down; but in other places they constitute forests 100 miles long and 10 to 15 wide.

They are almost as grand as the Big Trees themselves, being 50 to 70 feet in circumference, and 275 feet high; they form the entire forest (the Big Trees occurring in groups or groves among other trees), presenting therefore a grander sight, with their straight trunks without branches for 100 to 150 feet. The contrast of the cinnamon-colored trunks and the deep green foliage, shutting out the sunlight above, with the gloom and absolute silence of these majestic groves, prepares one to expect processions of ancient Druids emerging from these stately avenues, and to come upon some previously-undiscovered Stonehenge in these magnificent solitudes.

The groves of the "Big Trees" are found only between 36 deg. and 38½ deg. N. lat., and between 5,000 and 7,000 feet in vertical range. Of the eight or nine groves, the most famous are the Calaveras and Mariposa, the first the most northern of all. The Calaveras grove is the most accessible, and without horseback riding, and is distant only 74 miles from Stockton; of this distance, you may go nearly 30 by rail to Copperopolis, and the remainder by stages, riding directly into the grove, in which is situated the hotel. The grove is about the size of Boston Common, being about half a mile long and one-eighth of a mile wide, in a depression through which, in summer, runs a small brook. There are over a hundred large trees, 20 of which are more than 25 feet in diameter at the base, and many smaller, though very large ones. Some have fallen from age, and a few have been felled. The largest now standing, the "Mother of the Forest," is 320 feet high, 90 feet in circumference at the ground, and 61 feet in circumference six feet from the ground; the bark was removed up to a height of over 100 feet, and was exhibited in this country and in England, and was burned in the Sydenham Crystal Palace; there are pieces of it in this city more than two feet thick. The "Father of the Forest," prostrate on the ground, was the largest in the grove, estimated to have been 435 feet high, and 110 in circumference at the base; this is much larger than any now standing. One of the largest was felled in 1853 — 5 men working 25 days with pump augers and wedges; it was 300 feet high, and 96 feet in circumference on the ground; it was 80 feet in circumference 6 feet from the base, and large enough to accommodate four sets of quadrilles on the stump; and on its prostrate trunk, a house and double bowling-alley

80 feet long have been built. It was a section of this tree, cut 40 feet from the ground, that was exhibited in our Eastern cities last year;

this tree was probably not less than 1,300 years old. Another prostrate trunk, called the "Burnt Tree," will admit of a person on horse-

back riding through its hollow for 60 feet, in at one knot-hole and out at another. The tallest now standing is the "Keystone State," 325 feet high, but only 45 feet in circumference 6 feet from the ground; there are several others from 300 to 230 feet high, and 25 to 30 in circumference; and a large number still smaller, but splendid and symmetrical trees. The trees by which these are surrounded are so tall, that it is difficult to appreciate the height of these giants; when you reflect that the largest trees here are more than 100 feet higher than Bunker Hill Monument, and the "Father of the Forest" nearly as large at the base, you get some idea of their actual and relative size. The tops are almost always rugged and broken by the storms and winds, so that, as a general rule, they impress more by their size than their beauty and symmetry. When you are surrounded by trees 250 feet high, 50 feet more or less can hardly be appreciated by the eye. Other names of celebrated trees in this grove are "Hercules," "Hermit," "Old Bachelor," "Old Maid," "Siamese Twins," "Mother and Son," "Three Graces," "Gen. Jackson," "Daniel Webster," "Clay," "Washington," "Uncle Tom's Cabin."

The "Sentinels," about 50 feet in circumference, and 275 feet high, stand guard at the entrance of the grove, like giants at the portal of an enchanted palace; and between them, with head uncovered, you pass into this grand temple of nature.

The Mariposa Grove, about four miles southeast of Clark's Hotel, is also in a depression, accessible at present only on horseback or on foot. The grove is about two miles square, and its trees are more numerous, less lofty, but larger, than those of the Calaveras Grove. Many have names prominently affixed to them, taken chiefly from Americans famous in politics, science, literature, and especially poetry. Almost all are burned at the base, probably accidentally, by the Indians, and many have large cavities thus made in their standing trunks, through which you ride on horseback, and in which a large party could be protected from a storm. Many little trees

4

are growing all around, from two to one hundred feet high, and there seems no immediate danger of the species becoming extinct, especially as the groves are guarded and protected with the most jealous care by Mr. Galen Clark, the State Guardian. The first branches are given off at so great a height that it is difficult to obtain fresh specimens of the foliage; the cones are not more than two inches long, while those of the sugar pine, a large, but much smaller tree, are one and a half to two feet in length; the seeds are very small and light, and germinate readily in the East, and in Northern Europe; many are growing in this city and vicinity from seeds obtained and distributed by me last year; they grow with considerable rapidity, even two feet in a year, and form beautiful and interesting parlor ornaments. The foliage is somewhat like that of the arbor-vitæ; the bark smooth, porous, light, and cinnamon-colored; the wood red, as in redwood, light, spongy, and of not much use in carpentry. The largest tree in this grove is the "Grizzly Giant," ninety-three feet in circumference at the base, familiar to many by excellent stereoscopic views; the top is broken off, and it is evidently very old and declining.

THE CONE, AND FOLIAGE OF THE MAMMOTH TREES—FULL SIZE.

There are several other smaller groves, not generally visited by travellers. The species, therefore, can hardly be called a rare one, nor can it be said to be dying out. Though less high by one hundred feet than some Australian trees, and less in diameter than the African *Adansonia*, yet, taken altogether, it must be regarded as the grandest type of the vegetable world.

The white or bastard cedar (*Libocedrus*) resembles the big trees very much in its bark, and general appearance of the trunk; but the wood is white, and highly aromatic. Beside the large pitch or yellow pine (*P. ponderosa*), which here attains a very large size, the traveller will chiefly admire the sugar pine (*P. Lambertiana*), which grows to the height of three hundred feet, with a diameter of ten or twelve; this receives its name from a white manna-like exudation from the bark, whose sweet taste may tempt one to partake of it freely, to the great and painful disturbance of the abdominal contents, as it is a powerful purgative; the cones, of great size, hang like sugar-loaves from the branches. The traveller by the Mariposa route is generally taken to a large pine of this species, called the "Hermit's Cave," where an eccentric person passed a large part of the year; there was in its base, hollowed by fire, room enough for a bed of leaves, fire-place, and closets; the smoke of his fire ascended through a long chimney in the centre, the result of the natural decay of the tree. The dead branches of the pines are covered with beautiful bright yellow mosses and lichens, and the oaks in the valleys near the sea-level are festooned with long folds of grayish moss, which, swinging in the wind, give a funereal aspect much like that produced by a similar growth in the cypress swamps of the South. The dead and dying oaks display large mistletoes, three or four feet high, whose bright green forms a singular contrast to the ashy hue of the limbs at whose expense the parasite grows. One other characteristic tree deserves mention — the nut-pine (*P. edulis*), the seeds of which are largely eaten by the Indians; the wood of this tree is in great request for all kinds of structures under water; the wood is extremely crooked, and apt to warp in the air to such an extent that it is jocularly said that a "stick will crawl over a ten-acre lot in twenty-four hours."

### INDIAN TRIBES.

THE Shoshones, Utes, and Pah-Utes are the principal Indian tribes seen along the railroad from Salt Lake to Stockton. In the Yosemite Valley there are the " Diggers," so called because, in times of scarcity, they subsist on acorns, roots, and insects and their grubs, dug from the earth. Though low in the scale of man, they are not the abject creatures generally represented ; they are mild, harmless, and singularly honest. Of their honesty you can have no doubt when you see in the woods and valleys little storehouses, raised above inundations, and made of bushes, grasses, and stakes, in which their acorns and nuts are stored for the winter ; they always respect each other's property thus arranged, but these repositories have often been broken into and robbed by mischievous and unscrupulous whites. As usual with the American aborigines, they are more sinned against than sinning. They are very dark-colored, fond of gaudy beads and colors, and expert hunters and fishermen ; they will catch a string of trout where the Eastern angler, with his flies and costly outfit, cannot get a bite. They are addicted to intemperance, when they can get fire-water ; but for this, and the consequent poverty, misery, and disease, the whites are accountable.

While we were in the Valley, there was a grand pow-wow one night over the chief, who was supposed to be dying ; all sorts of howlings and incantations were practised by his women ; but the smell of his breath, his sudden revival at the mention of whiskey, and the fact that he was out fishing all next day, were sufficient proofs that it was only a fit of delirium tremens.

Near Clark's hotel is an Indian sweat-house, which is an object of curiosity to travellers. It consists of an oval depression in the ground, about eight feet long and two feet deep ; over this is a heavily-thatched dome-shaped roof, plastered with mud and leaves ; on the mud floor is placed a circle of rounded stones, enclosing a bed of twigs and leaves ; a fire is made around the stones, upon which, when highly heated, water is poured, at the same time extinguishing the fire, but raising an abundance of very hot steam ; the patient, naked, then lies down upon the inner bed of leaves, and the entrance is nearly closed ; after sweating sufficiently, he rushes out and plunges into a branch of the Merced River near by — a primitive but effectual Russian bath.

They possess the art of making baskets of straw which will hold water, and they make a very ingenious straw box for keeping their worm bait alive ; burying it in the earth, yet not allowing the worms to escape. The women are perfectly hideous, as usual doing all the drudgery, while the men hunt, fish, drink and smoke. One fine fellow at Mr. Clark's had charge of the train horses ; he was good-natured, strong, industrious, a fine rider, and skilled in all wood-craft.

It is averred by sundry persons not far from Cape Cod, that a baked skunk is a great luxury, and that, if properly killed and dressed, the flesh is not tainted with the well-known perfume of this animal. The Diggers are of the same opinion, and this dish with them corresponds to roast turkey with us. The following account of the manner in which the animal is captured by them is taken from a Western paper, and was written by an alleged eye-witness:

" On my journey hither, I observed two Digger Indians in a ravine, a little distance above the road, slowly and cautiously approaching each other, with their eyes intently fastened on some animal which a second glance discovered to be a well-developed specimen of the skunk. The Indian who was behind it held out his hand, and moved it slowly round in a circle, and this seemed to distract the attention of the animal, for he followed the motion closely with his eyes, and, though he elevated his tail several times, as if about to fire, he never executed his threat. Slowly, slowly they approached, the other attracted its attention, and the auspicious moment arrived. In the twinkling of an eye, the Indian behind dashed upon it, snatched it up by the extremity of its uplifted tail, and held it high aloft at arm's length. Then the other Indian ran up, flattened out his hand, and struck it on the back of the neck as he would have done with a knife, breaking that organ thereby, and the thing was accomplished. The animal seemed to feel itself so ignominiously disgraced and outraged, and all the proprieties and amenities of civilized warfare so utterly disregarded, in being hoisted by the tip of the tail, that it abandoned its usual means of defence in disgust. The consequence was that the entire operation was accomplished without the diffusion of the usual odor, which appears to be the main point in the killing."

The Mongolian origin of the American Indian has generally been accepted by closet ethnologists; but any one who takes this California trip will be likely to have this opinion, if he entertain it, shaken. Here you see the Indian and the Chinese side by side: except in the general contour of the face and the straight black hair, there is hardly any resemblance in physical character, and their mental characteristics are entirely opposite. The Diggers, and other California Indians, are supposed by some to have come from the west by sea, from the Japanese or Malayan Islands, instead of from the northeast, by way of Greenland, like the Esquimaux. Whatever their origin, they are fast disappearing, as they cannot adopt the civilization of the white race; scorning agriculture and manual labor, they are truly in the hunter state, and in their Stone Age, beyond which they will never progress.

## SAN FRANCISCO AND VICINITY.

IN and near the city of San Francisco, the traveller will find many fine scenes amid the Coast Range, even though fresh from the grandeur of the Yosemite and the higher Sierra. Within the city limits, by ascending Telegraph or Russian Hill on a clear day, you have before you a magnificent panorama; the splendid bay, dotted by sailing vessels and steamers from every clime, extending out to the vast Pacific through the Golden Gate — golden in the hues of an autumnal sun, and golden in the untold treasures to which it has afforded a pathway — the surrounding mountains, coming down to the sea, with their beautiful contrasts of reddish rock and green slopes, and their picturesque cañons rich in the trees characteristic of California — Alcatraz Island, with its fortifications, the more distant and lofty Angel Island — on the eastern side of the bay, the flourishing town of Oakland, noted for its University, and its connected villages, with the Contra Costa Range in the background, surmounted, though at a considerable distance, by Monte Diablo; to the south, from a neighboring hill, one may look into the San José Valley, famous for its mines of quicksilver; and many other objects crowd into the view, which the eyes must ever delight to look upon.

Monte Diablo, about 3,850 feet high, is very conspicuous, being quite isolated on the north, and its doubly-conical summit very graceful; it is distant from the city twenty-eight miles in a N. N. E. direction. The ascent is made from Clayton, which may be reached by land or by water; the distance to the top is only six miles, and may be easily made, and back, on foot or on horseback, in a day. The view from the summit is probably unsurpassed in extent, owing to the disposition of the mountains, and its position in the centre of a great elliptic basin. According to the geological survey of California, " the eye has full sweep over the slopes of the Sierra Nevada to its crest, from Lassen's Peak on the north to Mt. Whitney on the south, a distance of fully 325 miles. It is only in the clearest weather that the details of the 'Snowy Range' can be made out; but the nearer masses of the Coast Ranges, with their innumerable waves of mountains and wavelets of spurs, are visible from Mt. Hamilton (15 miles east of San José) and Mt. Oso on the south, to Mt. Helena on the north. The great interior valley of California — the plains of the Sacramento and San Joaquin — are spread out under the observer's feet like a map, and they seem illimitable in extent. The whole area thus embraced within the field of vision, as limited by the extreme points in the distance, is little less than 40,000 square miles, or almost as large as the whole State of New York." Extensive mines of bituminous coal have been opened here, and yield a large supply for the city.

The report continues : " What gives its peculiar character to the Coast Range scenery, is the delicate and beautiful carving of their masses by the aqueous erosion of the soft material of which they are composed, and which is made conspicuous by the general absence of forest and shrubby vegetation, except in the cañons, and along the crests of the ranges. The bareness of the slopes gives full play to the effects of light and shade caused by the varying and intricate contour of the surface. In the early spring, these slopes are of the most vivid green — the awakening to life of the vegetation of this region beginning just when the hills and valleys of the Eastern States are most deeply covered by snow. Spring here, in fact, commences with the end of summer; winter, there is none. Summer, blazing summer, tempered by the ocean fogs and ocean breezes, is followed by a long and delightful six months' spring, which, in its turn, passes almost instantaneously away at the approach of another summer. As soon as the dry season sets in, the herbage withers under the sun's rays, except in the deep cañons; the surface becomes first of a pale green, then of a light straw yellow, and finally of a rich russet-brown color, against which the dark-green foliage of the oaks and pines, unchanging during the summer, is deeply contrasted."

Among the many points of interest in the Coast Ranges, easily accessible from San Francisco, are Clear, and Borax Lakes, about 65 miles N. W. from Suisun Bay, and 36 miles from the coast. Borax Lake is a depression on the east side of the narrow arm of Clear Lake, from which it is separated by a low ridge of loose volcanic materials, consisting of scoriæ, obsidian and pumice. It varies in size according to the time of the year, and the comparative dryness of the season. In September, in ordinary seasons, the water occupies an area about 4,000 feet long and 1,800 feet wide in the widest part, irregularly oval, its longest axis being about east and west, with an average depth of 3 feet; it has been known to extend over twice this area, and has been at times entirely dry. The water from the lake contains about 2,400 grains of solid matter to the gallon, of which about one-fourth is borax. The borax, being the least soluble substance contained in the water, has, in course of time, crystallized out to a considerable extent, and now exists in the bottom of the lake in the form of distinct crystals of all sizes, from microscopic dimensions up to two or three inches in diameter. These crystals form a layer immediately under the water, mixed with blue mud of varying thickness. It is believed by those who have examined the bottom of this lake that several million pounds of borax may be obtained from it by means of movable cofferdams at a moderate expense. According to the San Francisco papers, during the year 1865 this lake supplied the local demand for borax to the amount of 40 tons, and yielded 200 tons additional for shipment to New York. It is collected from the mud at the bottom of the lake during the dry season, at the rate of about 2½ tons per day. The crude borax, thus obtained, is so pure, that the mint and assayers of the city use it in preference to the refined article brought from abroad.

In regard to the minerals of California, Prof. Whitney has reported that of the 65 elementary substances found in nature, so far as known to chemists, there are not 40 which have yet been proved to occur in California in mineral combination, and more than 20 elements are wanting on the Pacific coast. Of these a few are extremely rare, but the absence of some is surprising; fluorine, a substance of very general distribution in its abundant source, fluor spar, seems to be wanting in California, unless it exist in some of the micas. Taking the whole Pacific coast, from Alaska to Chili, the following facts appear: The small number of species, considering the extent of region as compared with other parts of the world; the remarkable absence of prominent silicates, especially the zeolites; the wide spread of the precious metals; the abundance of copper ores, and comparative absence of tin and lead; the similarity in the mineralized condition of the silver; the absence of fluor spar as vein-stone; no mineral species peculiar to the coast. Black oxide of manganese has recently been found in large quantities in a mine in the Coast Range, not far from the city of San Joaquin.

The quicksilver mines at New Almaden, California, are in one of the branch valleys of the San José, about twelve miles from the town of that name, and about sixty miles south of San Francisco. The ore is a sulphuret of mercury, and is found irregularly disseminated among beds of clay, slates and silicious strata, supposed to belong to the Silurian age; though rich specimens will yield sixty-seven per cent. of mercury, the average is about thirty per cent. The Indians had for a long time used this cinnabar as a pigment, and had excavated fifty or sixty feet into the mountain in search of it; in 1824 the Spaniards attempted to work the ore for silver, and afterward, in connection with the Mexicans and English, worked it successfully for quicksilver, the annual product being estimated at a million dollars. In 1858 the United States took possession, and the present workings are entered by an adit two hundred feet below the old excavations, extending about 1,500 feet into the hill; side galleries extend from this in the line of the deposit. This is a very interesting place to visit, and you may be rapidly carried, doubled up in a box, along a tramway very far into the bowels of the earth; the pitchy darkness, abominable smells and noises, and rapid rate at which you are whirled through passages, where a projecting elbow or head would be attended with dangerous consequences, give a sufficiently vivid practical illustration of some parts of Dante's Inferno. The simplicity and effectiveness of the smelting operations, by which the volatilized mercury is arrested, will excite the admiration of the visitor. Though the atmosphere of the mines is not unusually unwholesome, the men and the animals employed about the smelting works are subject to salivation, skin diseases, and the other attendants of mercurial poisoning. Other productive mines are also worked in this neighborhood. The product of California in quicksilver is annually more than two million pounds, against three and a half million at Almaden, in Spain, and one million at Idria, in Austria; most of the American quicksilver is carried to China.

## MINERAL SPRINGS AND GEYSERS.

NO one should leave California without visiting the mineral springs of Calistoga, and the Geysers. Calistoga is about sixty-four miles from San Francisco, by steamer twenty-four miles to Vallejo, thence by Napa Valley Railroad about forty more, via Napa, to the mineral springs, the most celebrated on the Pacific coast. The chief medicinal constituents are iron, magnesia, and sulphur, the temperature varying from boiling hot to icy cold. The vapor baths envelop the body like a hot robe, hence the name. The situation is one of the most charming in this delightful valley, and is appreciated by crowds of summer visitors, the greater part of whom pass onward to the "Geysers," twenty-two miles farther. The mildness of the climate renders it especially suitable for the culture of fruit, and some of the finest vineyards in this vicinity are in Napa Valley. It is essentially an agricultural community, and though there is a very extensive distillery for the manufacture of brandy, from the pure juice of the grape, in Calistoga, it is said that there is neither a policeman, doctor, or lawyer a permanent resident of the place. The fishing is fine, and in the surrounding woods may be found a great variety of game, from the plumed quail to the huge grizzly. This favorite resort for health and pleasure is within three and a half hours of San Francisco, and may be reached twice daily.

About five miles from these springs, on a small elevation, is a petrified forest. All along the Central Pacific Railroad in the Sierrra Nevada section, the traveller sees at the stations specimens, some very large and beautiful, of agatized, or silicified, or petrified wood; but here we find a forest, not buried in the ground, but exposed to view on the surface, though they are also met with at various depths in the soil. Within a radius of a mile are more than thirty of these fossil trees, the largest being twenty feet long and six feet in circumference; this trunk is prostrate, the roots being still below the surface, and is broken squarely across, and into several pieces, evidently silicified before it fell, the soil once surrounding it having been removed, probably by denudation from geological causes, and at a remote epoch; the hill upon which they are found is almost solid rock, conclusively showing the action of powerful denuding agencies. The wood is so hard that it will scratch glass, and in it are occasionally seen beautiful opaline spots. I do not know that the kind of tree has been accurately determined, though it is probably of some hard wood found now in this region. I have heard of other localities, near the line of railroad, both in California and Nevada, where similar petrified trees have been noticed.

The "Geysers" are in Sonoma County, twenty-two miles from Calistoga, by stage through NapaValley, and about nine hours' travel from San Francisco. They merit a visit not only for their medicinal properties, equal to those of Saratoga or Baden-Baden, but for their curious phenomena among the wildest and most picturesque scenery of the Coast Range. Along their course runs the Pluton or Sulphur Creek, stocked with fine trout, though in immediate proximity to troubled and diabolical looking waters. The waters found in the Geyser Cañon are alkaline, sulphurous, or acid, forming efficacious remedies for various cutaneous, rheumatic, and chronic diseases; some are icy cold, others boiling hot.

From Lieut. Davidson's account, the reader may form a good idea of the qualities of these waters. About seventy-five feet below the hotel, is the first spring of iron, sulphur and soda, with a temperature of seventy-three degrees Fahr.; going up the Geyser gulch you come to the tepid alum and iron spring, with a temperature of ninety-seven degrees, forming, in the course of twelve hours, a heavy iridescent incrustation of iron; within twenty feet of this is a spring of a temperature of eighty-eight degrees, containing ammonia, Epsom salts, magnesia, sulphur, and iron, yielding crystals of Epsom salt two inches long; higher up is a boiling spring of alum and sulphur, with a heat of 156 degrees, and near it, also, a hot black sulphur spring.

The following paragraphs are taken from Lieut. Davidson's account of these Geysers.

"As we wander over rock, heated ground, and thick deposits of sulphur, salts, ammonia, tartaric acid, magnesia, etc., we try our thermometer in the Geyser stream, a combination of every kind of medicated water, and find it rises up to 102 degress. The 'Witches' Cauldron' is over seven feet in diameter, of unknown depth. The contents are thrown up about two or three feet high, in a state of great ebullition, semi-liquid, blacker than ink, and contrasting with the volumes of vapor arising from it; temperature, 195 degrees. Opposite is a boiling alum spring, very strongly impregnated; temperature, 176 degrees. Within twelve feet is an intermittent scalding spring, from which issue streams and jets of boiling water. We have seen them ejected over fifteen feet. But the glory of all is the 'Steamboat Geyser,' resounding like a high-pressure seven-boiler boat blowing off steam, so heated as to be invisible until it is six feet from the mouth. Just above this the gulch divides; up the left or western one are many hot springs, but the 'Scalding Steam Iron Bath' is the most important; temperature, 183 degrees. One hundred and fifty feet above all apparent action we found a smooth, tenacious, plastic, beautiful clay; temperature 167 degrees. From this point you stand and overlook the ceaseless action, the roar, steam, groans, and bubbling of a hundred boiling medicated springs, while the steam ascends one hundred feet above them all. Following the usually-travelled path, we pass over the 'Mountain of Fire,' with its hundred orifices, thence through the 'Alkali Lake'; then we pass cauldrons of black, sulphurous, boiling water, some moving and spluttering with violent ebullition. One white sulphur spring we found quite clear, and up to the boiling point.

"On every foot of ground we had trodden the crystalline products of this unceasing chemical action abounded. Alum, magnesia, tartaric acid, Epsom salts, ammonia, nitre, iron, and sulphur abounded. At thousands of orifices you find hot, scalding steam escaping, and forming beautiful deposits of arrowy sulphur crystals. Our next visit carried us up the Pluton, on the north bank, past the 'Ovens,' hot with escaping steam, to the 'Eye-Water Boiling Spring,' celebrated for its remedial effects upon inflamed and weak eyes. Quite close to it is a very concentrated alum spring; temperature, 73 degrees. Higher up is a sweetish 'Iron and Soda Spring,' fifteen feet by eight; and twelve feet above is the 'Cold Soda and Iron Springs,' incrusted with iron, with a deposit of soda; strong, tonic, and inviting; temperature, 56 degrees. It is twelve feet by five, and affords a large supply. The Pluton, in the shade, was sixty-one degrees, with many fine pools for bathing, and above for trout-fishing.

"The 'Indian Springs' are nearly a mile down the cañon. The boiling water comes out clear as ice. This is the old medicated spring, where many a poor aborigine has been carried over the mountains to have the disease driven out of him by these powerful waters. On its outer wall runs a cold stream of pure water; temperature, 66 degrees; and another water impregnated with iron and alum; temperature, 68 degrees. It is beautifully and romantically situated.

"We have not mentioned a tithe of those you pass at every step in your explorations — nor one day nor one week will reveal them all to the inquirer. Do not suppose that desolation, fire, and brimstone reign supreme — one of the wonders of the place is that grass, shrubs, and huge trees should grow on its very edge, and even overhang, in many places, the seething cauldrons below. The most varied wood abounds around you — oaks, pines, sycamore, willow, alder, laurel, and madrone."

Bayard Taylor, describing his visit to the Geysers, says: "The scenery is finer than that of the lower Alps, and the place is a mine of future wealth, and of thorough rejuvenation." Of the Witches' Cauldron he writes: "A horrible mouth yawns in the black rock, belching forth tremendous volumes of sulphurous vapor. Approaching as near as we dare, and looking in, we see the black waters boiling in mad, pitiless fury, foaming around the sides of their prison. Its temperature, as approximately ascertained by Lieut. Davidson, is about five hundred degrees. An egg, dipped in and taken out, is boiled; and were a man to fall in, he would be reduced to broth in two minutes. From a hundred vent-holes, about fifty feet above our heads, the steam rushes in terrible jets. I have never beheld any scene so entirely infernal in its appearance. These tremendous steam-escapes are the most striking feature of the place. The wild, lonely grandeur of the valley, the contrast of its Eden slopes of turf and forest, with those ravines of Tartarus, charmed me completely, and I would willingly have passed weeks in exploring its recesses.

"A pure alum spring, reminding me of the rock-alum spring in Virginia, is a great resort for dyspeptics. In fact, the properties of all the famous watering-places seem to be here combined, and invite the sick to come and be healed."

Among the features of this region are the hills of crude sulphur for chemical manufactures, as gunpowder, sulphuric acid, etc., of which it is said half a million tons are annually consumed. The climate is unsurpassed for its salubrity. The Geysers may also be reached by steamer to Petaluma, thence by stages in ten or eleven hours; this route leads through Russian River Valley, and though longer and more fatiguing than the other, is very pleasant; it is well to go by Vallejo and return by Petaluma.

The religious spirit of the old Spanish Jesuits is perpetuated in the names of saints and of holy things given to many prominent places; such are San Francisco, San José, San Mateo, San Pablo, San Diego, San Joaquin, San Bernardino, San Antonio, San Quentin, Santa Barbara, Santa Clara, Santa Cruz, Sacramento, Los Angelos, etc. As these priests had a keen sense of the beautiful in nature, they selected for their missions the most delightful sites, which now afford to the traveller some of the most charming spots in California. Prominent among these is San José, well called "the Beautiful." The valley is very fertile, and the climate healthful; and the pueblo of San José, with the mission of Santa Clara, a few miles beyond, grew to be a very thriving place. It has increased rapidly since the Americans took possession, and is now celebrated for its wealth and refinement, for its excellent schools and fine public buildings. Horse-cars run in the principal street — the Alameda — which is flanked on each side by a fine row of willows, planted by the priests more than seventy years ago, now completely overshadowing the road to Santa Clara; three railroads now converge to this place, which is the centre of a large manufacturing interest; the population is estimated at over ten thousand. Santa Cruz, accessible by stage from Santa Clara, opposite Monterey, is a popular resort for excursionists, and is noted for its delightful climate.

California boasts, among other big things, that she has the largest orchard in the world. An English gentleman thus describes it. He says: A few days ago it was my good fortune and pleasure to visit an orchard located about two miles south of Yuba City, in Sutter County. The proprietor is the owner of 426 acres, mostly bottom land, lying along the west bank of the Feather River. The soil is a rich, sandy loam, and composed of the yearly deposits of the river many years ago. No better or richer land is to be found in the State. Before reaching the orchard proper we rode through a field of 150 acres of castor beans, growing in the most luxuriant manner—which field is to give place to a new orchard next year, the fruit-trees for the same at present growing in the nursery by the side of the field of castor beans, and containing 25,000 one-year-old budded peach-trees, 13,000 plum-trees, 6,000 Eastern walnuts, 25,000 California walnuts, 2,000 apple-trees, 500 Italian chestnut-trees, etc. Passing along through this forest of young trees, we arrived at the present peach orchard, consisting of 600 trees, two years old, and some of them bearing, this season, 150 pounds of peaches. These trees have made a remarkable growth, owing to the rich ground upon which they are planted, and in another year will make a tremendous yield of fruit. We next rode into the cherry orchard, containing 3,000 of the most

thrifty young trees ever seen on any ground,  The different varieties,
fifteen in number, gave this orchard a variety of aspect, and broke
up the usual monotony of the steeple-like formed cherry orchard.
These cherry-trees were all imported from Rochester, N. Y., about
three years ago.  Off to the south of this wonderful wilderness are
2,000 plum-trees, of twelve varieties, and 500 apple-trees, mostly
winter varieties.  Passing the peach orchard we reached the apricots,
2,000 in number, which are also two years old, and have borne a fair
crop the present season.  This is really a California wonder.

## HOMEWARD BOUND.

THE traveller, having visited the above places in the vicinity of San Francisco, will now think of turning his face eastward, if he return overland, and of examining more closely some of the interesting points which he hurried by in his eagerness to behold the wonders of the Yosemite. The first place which will claim his attention is Oakland, so called from its beautiful groves of oaks, opposite San Francisco, and fronting the Golden Gate. The shallowness of the water in the bay has compelled the railroad company to build a wharf about two miles long into the bay, so that you seem to be going out to sea in a railroad car; from the end of this wharf is established the ferry to San Francisco, being the terminus of the Western Pacific Railroad. Sometimes called the "Park City," it bears somewhat the same relation to San Francisco that Brooklyn does to New York; it is, *par excellence*, the educational centre of California; besides its numerous public and private schools for both sexes, being the site of the State University. The drives along its macadamized streets, with the fine view of the bay and the distant Pacific, and the beautiful gardens on every side, cannot be surpassed, if equalled, in any city of the country.

One of the most interesting cities which the Yosemite tourist is sure to visit is Stockton, about ninety miles from San Francisco by railroad. It was named in honor of Commodore Stockton, who took an active part in the conquest of California, and was laid out by Capt. Webber in 1849–50; it is also at the head of navigation on the San Joaquin River, distant by water 127 miles from San Francisco, and accessible by large steamers and sailing vessels; the river is navigable for small steamers more than 100 miles farther up. It is estimated to contain about 12,000 inhabitants, and is a very busy and thriving place. The public and private buildings and stores, many of which are built of brick, give it a decidedly Eastern look. Near the Yosemite hotel, the principal one, is the enclosure which contains the State Asylum for the Insane. The country around Stockton is exceedingly fertile, and its agricultural resources are inexhaustible; its mining facilities are also important. An artesian well, 1,000 feet deep, supplies the city daily with 360,000 gallons of water; though the water rises eleven feet above the surface, it is raised by steam to a high reservoir, whence the city is supplied. It is in the centre of the vast grain-producing district of the San Joaquin Valley; and in harvest time the roads are lined with the mule-drawn wagons heavily laden with the golden produce, which has been estimated at $3,000,000 annually. The soil around the city is a black vegetable mould, called "adobe," soft and slippery in the rainy season, hard and deeply cracked in the summer; about five miles beyond this begin the sandy plains leading to the foot hills, described in a previous chapter.

Stockton is well called the "Windmill City," as, by sinking a well-tube ten to twenty feet, water is readily obtained. Hence almost every one cultivates the rich soil as a garden, watering it by his wind-pump, which takes the place of the hand-pump in almost every yard. The gardens are very beautiful; and, such is the mildness of the climate, figs, and other sub-tropical plants, flourish and ripen in the open air. This is the centre of the stage lines for the Yosemite Valley, and both the starting and return point for most travellers bound for that region. In the summer season, when the water is low, the sloughs which penetrate the city in various directions have a green, stagnant, and most unwholesome look; they receive much of the drainage of the houses, and cannot fail, sooner or later, to form a suitable receptacle for the origin and spread of epidemic disease, when drought, heat, and accumulation of filth shall unfortunately occur together.

Leaving San Francisco at 8 A. M., you reach, on your return-trip to the east, at about 5 P. M., the pretty and flourishing town of Colfax, 192 miles, named from Vice-President Colfax. Here it is well for those interested in mines to stop a day or two to pay a visit to Grass Valley and Nevada, among the most important of the gold-producing regions of California. Grass Valley was one of the earliest stopping-places of the old "forty-niners," not only because there they found excellent pasturage for their animals, but on account of the profitable "washings" from the streams; the subsequent discovery of rich veins of gold-bearing quartz led to the building up of a town, numbering now about five thousand inhabitants. The fine orchards and gardens around the miners' houses render this one of the prettiest of the mining localities, and show that the thirst for gold does not necessarily interfere with the love of the ornamental and the beautiful. Its buildings, newspapers, schools, and churches, distinguish it as a centre of enterprise, intelligence, and wealth; there is probably no place in the State where mining improvements and machinery are better appreciated, and more successfully employed, than here. It is thirteen miles north of Colfax, and easily accessible by a line of stages. Though about 2,600 feet above the sea, it is so far below the snow-line, that its temperature permits the ripening of semi-tropical fruits, and its climate is very healthy.

Nevada, four miles distant, the county seat, can also boast of very fine buildings, and a considerable population engaged in mining and agriculture; it is rather irregularly laid out on both sides of Deer Creek, which runs through a part of the town. After the washings in the old river had ceased to be profitable, hydraulic mining was introduced with great success; but now the principal mining operations are upon the quartz in the fine stamp mills. It has been estimated that over fifty million dollars' worth of gold has been taken from this locality in twenty years. Newspapers, banks, churches, and schools, indicate the prosperity of the place. A foundry, flouring-mills, and distilleries, show that manufactories and agriculture may be profitably pursued in busy mining regions; the soil of the valley and surrounding hills is well adapted to the fruits and vegetables, which are the pride and boast of California, and the delight of the hungry traveller.

Passing eastward 65 miles from Colfax, you come to Truckee, a large, busy, and muddy town, of over 4,000 inhabitants, chiefly engaged in the lumber business; it is situated in a heavily-timbered region.   The traveller would make no stop here, were it not the starting-point for Lakes Tahoe and Donner, which are indeed the gems of the Sierras.   The Truckee River, which runs along the road for miles, brawling in its rocky bed, has one source in each of the above lakes, and empties its waters into Pyramid Lake to the north.

Lake Tahoe is 12 miles distant, and the road along the river bank is delightful.   The dividing line between California and Nevada runs through the lake, and its waters wash the shore of five counties; the depth along this line is about 1,700 feet.   No words can do justice to the beauties of this lake, before which those of Como and Maggiore are not to be mentioned; the crystal purity of the water, the mountain slopes, the verdant meadows, the splendid trees, to say nothing of the pleasures of sailing, fishing, and shooting in its invigorating air, excuse the raptures into which every appreciative traveller involuntarily falls.

Donner Lake, much smaller and deeper, and equally beautiful, and always memorable from the terrible event which has given it its name, is only two and a half miles north-west of Truckee.   Both these lakes are noted for their silver trout, which attain the weight of 20 pounds, and test the skill of the angler to the utmost.

This brings us to the confines of California, to go beyond which is foreign to the purpose of these pages; the most noteworthy points on the return east are the famous Comstock and other silver lodes, at Virginia City, Nevada, whose wealth is almost incalculable, and the Shoshone Falls, in Idaho, over 200 feet high, and said to exceed Niagara in the grandeur and wildness of the surrounding scenery, though with much less volume of water.

Then you may leave the Pacific road at Cheyenne, and go south to Denver, and from that point spend a few weeks most profitably in exploring the magnificent scenery of the parks of Colorado.

Some travellers, having a love of the ocean, and plenty of time at their disposal, may prefer, as I did, to return once by sea from San Francisco, *via* Panama and Aspinwall; for what may be enjoyed on this trip, the reader is referred to the next chapter.

## SAN FRANCISCO TO BOSTON.

ON taking the ferry-boat at Oakland to make the six or eight miles' transit across the bay to San Francisco, I was surprised to find the ladies dressed in furs, and the gentlemen with winter overcoats; the air was damp and chilly, very much like a Boston east wind in March. From April to November, the ascent of the heated air from the valley of the Sacramento along the Coast Mountains to the east causes the cold north-west·winds to rush in from the Pacific through the Golden Gate, laden with moisture, whose condensation envelops the city in the morning and evening in dense fogs, with many clouds, which never at this season yield any rain. The hot sun at mid-day dispels the mists, and straw hats and thin garments are worn at noon of a day whose morning temperature was disagreeably cold. This season is admitted to be the most uncomfortable in the whole year, and the most trying to invalids. The same wind which blows up the clouds of sand in the streets, roughens the waters of the bay, and makes the passage in or out rather cold and dismal. Soon after getting out of the Golden Gate and on to the Pacific, the wind dies away and the sea becomes smoother, but the clouds without rain, and the cold fogs, accompany you for hundreds of miles at this season (August). The rocky islands and headlands give shelter to innumerable sea-birds, especially guillemots (*Uria*), whose large and irregularly blotched eggs are sold by the hundred for food in the San Francisco markets; there are also many large seals, or so-called sea-lions (*Phoca jubata*), about the same rocks. This cold, damp, and foggy air does not go very far inland; and in the foot-hills, and higher mountains, the sky is cloudless, the nights without dew, and the stars as bright as on a frosty night with us; the air is so dry that there is no danger of taking cold in camping out, even at an elevation of five thousand feet; and travellers not unfrequently place their cot-beds on the outside and uncovered piazza, sure of a pure, dry air, with no danger of rain; it is this rest you get at night, which enables you to rise refreshed after the heat, dryness, and dustiness of the day's travel.

One of the striking characteristics of the Pacific steamers is, that the crew are all Chinamen; and any one who has experienced the disorder, the dirtiness, the unnecessary noise, scoldings, swearings, and often intoxication, attendant on the sailing of ships from Atlantic ports, must be delighted with these Chinese sailors; they are neat, orderly, quiet—not using oaths, tobacco, nor whiskey—obedient, respectful, strong, and in every way good sailors.

The coast, seen at a distance of about three miles, is high, rocky or sandy, but indescribably barren and inhospitable looking. The sea, for the whole voyage of two weeks, was remarkably smooth, well justifying the term Pacific to any one who has been tossed about on the Atlantic; except in crossing the gulf of California, there was no more roughness, exclusive of the long and gentle tidal swell of the ocean, than an hour's east wind would create in our bay. In fact this now rarely undertaken Pacific voyage is, at this season, very delight-ful, with its beauty, and quiet, and absolute repose of body and of mind, fully realizing the dreamy *dolce far niente* of the Italian imagination. Large petrels (*Puffinus cinereus* — Gmel.) began to appear and fol-low us on the second day out. On alighting in the water, which they often do, they put forward their webbed feet, checking their headway in this manner, backing water as it were, with the wings spread, be-fore settling on the surface. They came around and near the steamer in considerable numbers, but never alighted on it, as the booby of the Atlantic does. On account of the great length of their wings, and the shortness of their legs, they cannot rise, like the gulls, directly from the water, but are obliged to run along the surface, like the smaller petrels, beating the water with their feet, until sufficiently elevated to use their wings.

Flying fish also began to appear, but neither so numerous, nor so large, as in the Southern Atlantic. The ventrals were expanded just like the pectorals in the act of flight, the former being much the smaller. They rose out of a perfectly smooth sea, showing that they are not mere skippers from the top of one wave to another; they could be seen to change their course, as well as to rise and fall, not unfrequently touching the longer lower lobe of the tail to the surface, and again rising as if they used the tail as a powerful spring. While the ventrals may act chiefly as a parachute, it seems as if the pectorals performed, by their almost imperceptible but rapid vibrations, the function of true flight. Another reason which leads me to think they perform a true flight, is the way in which they reënter the water. After reaching the end of their aërial course, they drop into the water with a splash, instead of making a gentle and gradual descent, like the flying squirrel, flying dragon, and other vertebrates with mem-branes acting as parachutes. The drying of the flying membrane in the air would prevent the small but numerous and rapid motions nec-essary for true flight, and the animal therefore suddenly drops when the membrane becomes stiff. I do not see how the drying of the pectorals would affect their action as parachutes. The temperature of the air was 70 deg. Fah.

At the same time there were seen small Portuguese men-of-war (*Physalia*), no larger than an olive, and without the purple reflec-tions of the larger ones so often met in the Atlantic. Whether these were the young or full-grown individuals I do not know; I saw none larger than these, and they were not numerous.

As we approached the coast of the gulf of California the petrels left us, and were replaced in an hour or two by white gulls about the size of Bonaparte's gull, but either entirely white, or with a very slight lavender-blue tinge on the back and wings. These had an en-tirely different way of alighting, and rising from the water; they did not put forward their feet to arrest their course, but circled round like pigeons until their headway was stopped, and then quietly set-tled upon the water, immediately folding their wings. They also rose directly from the surface, without running along as the larger-winged petrels did.    75 deg. Fah.

The next day, August 7, the temperature was 80 deg. Fah. Land was in sight all day. The California coast, for hundreds of miles, is most forbidding, rocky to the ocean, with high mountains in the background, entirely parched and barren at this season, and having that greenish-red tinge suggestive of mineral contents, especially copper. The shore is entirely uninhabited even to beyond the mountains, and shipwrecked persons there would perish of starvation if they depended on what the country afforded. Indeed a part of the coast near which the "Golden City" went ashore in 1869, is called "Starvation Point"; her numerous passengers, among whom were many women and children, had to walk more than twenty miles to reach a headland, where their signals of distress were fortunately seen by a passenger on one of the Pacific steamers bound in the opposite direction, who was trying his opera-glass very early on that morning. There is now little commerce in these waters, and we did not see a sail for days on this part of the coast; all the trade is done by a few small coasting schooners, which keep near the shore. The coasts of Mexico, Nicaragua, and Costa Rica, on the contrary, are beautifully green.

After passing Cape St. Lucas, August 8, we were in the mouth of the Gulf of California, where it ascends many hundred miles to the north, parallel to the coast, leaving the long, comparatively narrow, barren and uninhabited region, along which we had sailed for the past two days. The weather now became hot — 85 deg. Fah. at noon, and so remaining day and night to Panama, once going up to 88 deg., and occasionally descending to 84 deg. Point Conception, in latitude 34 deg. 50 min., corresponds very nearly to Cape Hatteras on the Atlantic coast; at this point, the coast, instead of continuing to follow the mountains from north-west to south-east, becomes nearly east and west, and the cold north-west winds from San Francisco are suddenly exchanged for the warm southerly winds of the tropics, and off goes the pea-jacket, and on goes the thin coat and light hat. For two or three nights, the nearly full moon shining upon the glassy sea was very beautiful; but with the moon, as with the sunrise and sunset, I find that we have far more beautiful colors and contrasts at home; it seems as if the land and sea must be both before the sight to give the full effect, which a dreary waste of water alone cannot give.

The water here was very phosphorescent. I obtained a bottleful in about latitude 19 deg., which has been unopened since August 9. It may be interesting to see if it contains more salt than the water of the Northern and the Atlantic Oceans, as is alleged — if there be in it any remains of diatoms, or of animal forms, or of any kind of organic or nitrogenous matter which may serve as nutriment for protozoa, or any dilute protoplasm diffused through the waters of the ocean which could be directly absorbed by these lowest organisms.

The Mexican shore here came in sight, strikingly contrasting with the Californian, being green, with a luxuriant vegetation, and very pleasant looking; the shore high, with elevated mountains in the distance, and here and there a beach lined with coral reefs against which the surf could be seen breaking. We could see the rain-clouds

in the mountains, and the lightning, and hear the thunder; while
where we were — three miles from the shore — all was bright sunshine,
with no sign of rain.  On the ninth, in about 18 deg., we stopped in
the land-locked harbor of Manzanillo, the mountains rising steeply
from the water's edge, more than one thousand feet high, clothed with
vegetation to the very top.  For the last day, after leaving the California gulf, no birds were seen; first we had the large petrels, then
the smaller white gulls; these soon disappeared, having limits beyond
which they did not pass; the reason was not evident to our senses,
as the climate, and the shore, and the sea, appeared to us the same;
but the birds knew the difference.

On the eleventh we reached Acapulco, Mexico, in about 17 deg.
north, where we stopped half a day, going on shore to purchase shells
and corals, and the luscious fruits of the place, and to witness the
strangeness of an old Mexican city, with its Spanish decay softened
by tropical indolence, its curious mixture of natives, negroes, and
Mexicans, the peculiar customs of the market-place, and the heter-
ogeneous articles exposed for sale; the stock of a hundred women,
and nearly as many men, was not greater than the contents of a single
stall in one of our markets, the trade being of the most petty description, and seemingly like that of children playing buying and selling
merely to pass away the time.  I obtained here a few shells, especially
*murices*, and some natural and artificially-colored corals.  The harbor
is very beautiful, entirely land-locked, surrounded by high hills covered with bushes to the top; here and there could be seen the palm-
leaf huts of the natives, with patches of bananas and groves of oranges;
the beach was lined with palm-trees, and everything had the peaceful,
lazy, dreamy look peculiar to the tropics; the buildings of the town
are of stone, with tile roofs, and generally of one story; the old
church in the plaza was built by the Spaniards, and is now used as a
prison, as its grated windows indicated.  The water was beautifully
clear, and swarmed with bright-colored fish, and it is said with sharks;
I saw none of the latter, and the professional divers near the landing
apparently had little fear of them, as they dived for the pieces of
money thrown to them by the passengers.

When the coasts of Southern Mexico and Guatemala are reached, and
especially about latitude 11 deg. 30 min., white-rumped Mother
Carey's chickens came around us; they looked just like the common
Atlantic species, and, as Baird does not describe such a bird on the
Pacific in vol. ix. of the Pacific Railroad Reports, I suppose the
species must have appeared since then, either from South America, or
having crossed the isthmus.  Now and then a marine turtle would be
seen lazily rolling at the surface.

The lowest latitude reached, is about 7 deg. north.  We arrived at
Panama Aug. 17 (a fortnight from San Francisco), where we re-
mained two days, giving ample time to examine this quaint old Spanish town.  In the spacious and fine harbor were many hooded gulls,
brown pelicans, and frigate pelicans, while numerous turkey buzzards
ran along the beach with the same tameness and voracity as in our
Southern and Gulf States; the water abounds in catfish and sharks,

though I saw none of the latter caught by the numerous fishermen. Panama is built along the bay, which is surrounded by high hills and mountains, covered with tropical verdure; many of the smaller islands show columns of basalt with precipitous sides, and there have been several noted subsidences of the land. Though hot in mid-day, the temperature at night was delightful; and this in the middle of August. The place has the typical appearance of a dirty Spanish town.

We left Panama, Aug. 19, to cross the Isthmus to Aspinwall, a distance of forty-seven miles, occupying three hours in the passage, in very dirty and uncomfortable cars, steerage mingled with cabin passengers, as both classes pay the same fare, viz., twenty dollars in gold. The route runs for nearly half the distance along the Chagres River, a narrow, muddy stream, with banks of reddish clay which tinges the water to the color of that of the Missouri River; the road has some sharp curves, and a few cuts, and presents only one engineering notability, where it crosses the river on a substantial iron bridge. The land is mostly low, and the vegetation most luxuriant; water seems abundant, but of a repulsive look and stagnant character, which, with the marshy effluvia, fully explains the death of thousands from malarious disease during the construction of the road; it is familiarly said that a life was lost for every sleeper laid, so unhealthy was the region for Northern workmen. The natives, however, seemed vigorous and well developed, and every hut swarmed with children, the amount of clothing on which, especially on boys to the age of seven or eight years, would not materially draw upon the contents of a dry goods store. Many negroes were seen, and they fraternize fully with the Indian natives; the latter are nearly as dark as negroes, but have finer forms, more regular features, and straight black hair. The marshes and the mud are occasionally relieved by masses of very dark volcanic looking rock, through which several cuts have been made; the graceful palms, and the beautiful flowers, could not fail to attract the attention of the most unobservant; the only birds seen were small black anis (*Crotophaga ani.* L.), a scansorial bird of the cuckoo family, which hopped and flew about like blackbirds with us.

The town of Aspinwall is small, low, on the margin of a swamp, recalling to the mind the ideal of the marshes of the carboniferous period, and suggesting the formation of coal from the luxuriant vegetation; though, near the sea, the water is salt, instead of the fresh water supposed to be necessary to the formation of coal.

There was nothing noteworthy in the nine days' passage to New York, except the much greater heat in the Caribbean Sea, than in similar latitudes on the Pacific; probably from its comparatively small size, and being land-locked. No whales were seen in the Pacific, and none in the Atlantic, till latitude 37 deg., off Delaware Bay, when a school of about twenty finbacks, some of them forty to fifty feet long, came quite near the steamer; I was interested to notice that their blowing projected into the air simply a fine vapor, and not a jet of water, as is usually believed; that cetaceans do, however, sometimes eject water in this way, I know, as I have, on many occasions, at night, heard the puff soon followed by the swash of the descending water.

The whole trip from San Francisco to New York takes about twenty-three days, at a cost of $100 in gold; in the cars you can make the passage in one-third the time (seven days) at a cost of about $180 — by the cars, two weeks shorter and about $60 dearer — if one has plenty of time, it is far pleasanter by sea, as you are brought into contact with new aspects of nature, tropical scenery and fruits, and are free from dust, change of cars, anxiety about baggage and sleeping facilities, and from the inevitable rush of the dining saloons and railway stations.

In these short sketches I have endeavored to express what especially interested me in the California trip; others will take note of different things, each according to his taste and education; but every one will, I think, admit that this journey will bring him into contact with some of the sublimest of scenery.

As to the causes which have produced this remarkable Valley, there are three principal theories: the subsidence theory, the ice theory, and the water theory.   From what I have seen, and have been able to ascertain, it seems to me that there was a great subsidence, as claimed by Prof. Whitney, and that subsequently an immense glacier extended to the edge of the Valley, even entering the westerly end of it by the numerous cañons there, as proved by the glacial scratches and moraines, and giving rise, by its melting, to a great lake, which gradually disappeared.   That the Half Dome, El Capitan, and other masses in the Valley, were produced, or essentially modified by ice or water, I am not, with the present evidence, prepared to believe.

As a means of restoring impaired health, and of invigorating the feeble and nervous of both sexes, it is to be highly recommended — its bracing air, pure water, delightful tramps, and awe-inspiring scenery, are a thousand times more to be desired by persons of sense and culture, than the inanities of Saratoga, the fashion of Newport, the pomposity of Long Branch, the petty swindling of Niagara, or the discomforts of the White Mountains.

# INDEX.

## PHOTOGRAPHIC VIEWS.

www.ingramcontent.com/pod-product-compliance
Lightning Source LLC
Chambersburg PA
CBHW022011190326
41519CB00010B/1475